Interstellar Tours

Interstellar Tours

A Guide to the Universe from Your Starship Window

Brian Clegg

ICON

Published in the UK and USA in 2023 by
Icon Books Ltd, Omnibus Business Centre,
39–41 North Road, London N7 9DP
email: info@iconbooks.com
www.iconbooks.com

ISBN: 978-183773-075-9
ebook: 978-183773-077-3

Typesetting by SJmagic DESIGN SERVICES, India

Printed and bound in the UK

For Gillian, Rebecca and Chelsea

CONTENTS

ACKNOWLEDGEMENTS

Thanks, as always, to the excellent team at Icon, especially Duncan Heath.

I'd also like to remember those who got me interested in space and astronomy, many now sadly no longer with us. These include my father, Leonard Clegg, who drove with me halfway across the country to pick up the six-inch reflecting telescope that was my first experience of anything other than naked-eye astronomy. Then there was Patrick Moore for his eccentric but engaging commentaries on the long-running *The Sky at Night* TV show (and for his supportive answer to a letter of mine). Not to mention our only real space travellers to date, the astronauts who made it to the Moon, from Neil Armstrong to Eugene Cernan, who at the time of writing was the last person to have walked on the Moon – although hopefully this will not be the case for much longer.

It's easy to think of space travel as science fiction – and it's certainly the case that science fiction enables us to explore the universe (or, indeed, a whole range of universes) in our imagination. But I would like to hope that at some time in the future, even if not on as grand a scale as is possible for readers of this book, human space travel to the planets of our solar system and beyond will be more than just a dream.

PRE-FLIGHT CHECKS

A few years ago, I wrote a book called *Inflight Science* that described the science that we experience when on a plane flight, giving the reader a chance to explore the things we observe through the airplane window as well as the science and technology that makes the flight itself possible. *Interstellar Tours* is a sequel to *Inflight Science* that takes things to the next level. This is the science we might experience on an interstellar space liner of the 22nd century.

As much as possible, the facts are as solid as those featured in the pages of *Inflight Science*. However, cosmology is inevitably a very different science from the physics of our world. It features more speculation because we can't do experiments or directly examine many of the phenomena that we are looking at from a vast distance. Where that's the case, I will make this obvious in the text, but what we will experience on the flight is based on our current best understanding.

The only exception to this approach is in the first full chapter, Welcome Onboard. With our current technology, it is not possible to make a journey to distant stars. To be able to take our journey across the galaxy, we will discover the need to make use of technology that is pure science fiction after exploring and dismissing hypothetical technology that is based on current theory. Certainly, for the moment, there is no feasible way to make human interstellar flight practical.

Science fiction is sometimes treated as second-class literature – but I think it gives us a wonderful opportunity to ask, 'What if?' In recent years, the genre has been caricatured and denigrated, with Margaret Attwood even dismissing it as 'talking squids in outer space'. But the real science fiction is about telling stories that explore the ways that humans might react to the impact of new science and technology on their lives and their environment. Although this book is set in outer space, there will be no talking squids – its role is to bring us closer to astronomical and cosmological features that would be on the tourist trail if only we could get out there – all based on the best current science.

Given that we won't be meeting any talking squids, it's worth reflecting why this is the case. Around 1950, the Italian physicist Enrico Fermi was taking part in a discussion around a canteen table at Los Alamos, the site of the US atomic bomb project during the Second World War. The subject of flying saucers (as Unidentified Flying Objects or UFOs were then known) came up. Fermi is reported to have said to his colleagues, 'Where is everybody?' It seemed strange that there was so much talk of aliens from space but that there was no scientific evidence for the existence of alien life.

For centuries there has been speculation about living beings (often all too predictably humanoid) existing on heavenly bodies. In the early days, this even included the Sun. And, since the 1940s, there was a steady stream of sightings of UFOs, or as we are now supposed to call them, unexplained aerial phenomena – UAPs. (They were renamed partly due to the US military's love of novel abbreviations and, more reasonably, because quite a lot of these phenomena are not flying objects.) One of those present at the Los Alamos discussion remembers seeing a cartoon at the time explaining that the disappearance of trash cans from New York City was the work of little green men who had been taking them away. Aliens were very much part of the zeitgeist. But they were hardly knocking at the door of the White House.

Although we can't be certain, there is no credible evidence that UAPs have any connection with visitors from other planets. As has been widely pointed out, given that so many individuals now carry high-quality cameras with them at all times, built into their phones, it is strange that we still only get vague fuzzy photographs to support claims of UAP sightings. More so than ever, we can now ask, 'Where is everybody?'

It seems unlikely, then, that we have been or will be visited by aliens. There certainly may be alien life out in the galaxy (though if there is, the chances are that the majority of it is more similar to bacteria than humans), but we don't have any good scientific evidence to back up its existence. The old saying 'the absence of evidence is not evidence of absence' is technically true – but it is simply telling us that, for now, science, which needs to be evidence-based, has nothing useful whatsoever to say about alien life. All we have is pure speculation, unsupported by evidence, or stories.

I'm not knocking aliens in science fiction – there are some great characters, though the ones we develop affection for, from Mr Spock to Yoda, tend to be far more anthropomorphic than anything we might expect real aliens to be like. The reality is that fictional aliens are often playing a role that helps us explore what it is (and what it isn't) to be human. But until we have any scientific evidence, fiction is where they belong. That being the case, aliens will not be making an appearance on our interstellar tour.

There may be no aliens on our journey into outer space, but what you will be able to see is an impressive collection of images, provided on the *Interstellar Tours* website: www.interstellartours.co.uk. Where possible, these are photographic, but any featuring fine detail of distant systems, planets and solar system will be artists' impressions. The aim is for them to give you a sense of what it is like to experience a visit to these locations, but it is important to bear in mind that the exact detail shown may not be correct.

For the moment, though, it's time to take our place on board the starship *Endurance** and to get a quick introduction to the technology that is going to be necessary to make our voyage possible, and survivable.

---

* Our starship is named after Ernest Shackleton's wooden ship, lost in 1915 when it became trapped in pack ice when exploring Antarctica. We can never overemphasise the influence of the 20th-century TV show *Star Trek* on the interstellar touring business. Our entire starship fleet has three syllable names beginning with E.

WELCOME ONBOARD

1

One thing that the classic TV show *Star Trek* (mostly) got right is that starships don't land on planets. It's easy to underestimate just how difficult it is to get a massive object off the surface of a planet and into outer space. The problem lies in escaping the gravity well. An object as big and heavy as the Earth – which has a mass of around 6×10^{24} kilograms (13×10^{24} pounds) – holds on to objects on its surface with an iron grip. Even when birds or planes do make it into the sky, they soon have to return to the surface. What goes up really does usually come down.

> ### Units and stuff
>
> Occasionally we will be using scientific notation like the 6×10^{24} above. This is just a convenient way of representing large numbers. Here, 6×10^{24} is shorthand for '6 multiplied by 10, 24 times over' – or to put it another way, 6 followed by 24 zeroes. You could also say that it's 6 trillion trillion.
>
> Science makes use of the metric system for all measurements, and, by the time the *Endurance* was commissioned, no one on Earth was still using the traditional units such as feet or pounds. They had gone the same way as rods, poles, perches,

bushels and chains as units of measurement. However, for old times' sake, we will show both metric and traditional 'Imperial' units, except for restricting weights to metric tonnes, as these are close enough to traditional tons to make the distinction unnecessary.

We are used to measuring long distances in kilometres or miles, but in space, a kilometre is a pathetically small unit. The most useful measure for us will be the light year – the distance that light travels through space in a year. A light year is 9.46×10^{12} kilometres or 5.88×10^{12} miles. To put that in context, the distance from the Earth to the Sun is about 8.3 light minutes or 0.000016 light years. Astronomers often prefer to use distance units called parsecs (which are around 3.26 light years). These work particularly well with the mechanics of telescopic observations, but we will stick to light years as they are easier to envisage.

Are you massive?

Because we are going to spend our time during the journey out in space, it's worth quickly clearing up the distinction between mass and weight, because the difference matters very much when you are away from the surface of the Earth. These terms tend to be used interchangeably back home, but they are very different things, and in space this will become obvious.

Mass is an intrinsic property of an object, which is measured in kilograms (officially, the traditional mass unit is called a slug (14.59 kilograms), although the pound tends to be used more often). It doesn't matter where an object is, it will always have the same mass, unless bits are removed from it or added to it. You could see mass (a concept introduced to the world by Isaac Newton back in 1687) as a measure of the amount of stuff in an object – whether that object is you, a starship or something as large as the Earth.

Weight, by contrast, is the force that is felt by an object under the gravitational pull of a body such as a planet. When we talk about the weight of something, we really mean 'its weight when it is on the surface of the Earth', though we tend to omit the last bit. Your weight would be totally different if you were on the surface of the Moon, for example – about a sixth of what it is on Earth. In space, your weight could be zero, though, as we will discover, it certainly doesn't have to be, and it will only be zero on the *Endurance* when in a special, low-gravity entertainment area. Having weight makes doing many things much easier – from eating to visiting the toilet.

Although your bathroom scales will give you your weight in kilograms or pounds or stones, this is a cheat. Strictly speaking, weight is the force due to gravity acting on the mass of an object. Scientifically, this should be measured in units called newtons (foot-poundals for traditional unit fans), but in practice we tend to fudge it and still use the mass based on what's measured on the Earth's surface. So, when we say that that on the Moon you will weigh one-sixth of what you do on Earth, what this means is that you will feel the force pulling you down that you would experience if you had one-sixth of your mass on the Earth's surface.

Whether we talk of mass or weight, the first stage of taking our interstellar tour is getting off the Earth. And here things have not moved on as much since the earliest days of space flight as 21st-century people might have expected. The old rockets were extremely unsafe and scary. Nonetheless, we are still using the equivalent of rockets, although with far less risky sources of propulsion, and the ability to make the journey into orbit under a level of acceleration that won't put the stresses on the body experienced by early astronauts. For us, it's no different from taking a plane trip. We have moved on from the first journeys into space, but not as much as science fiction writers might have hoped for.

As for the *Endurance* itself, nothing as massive as a starship could survive landing or take-off from a planet. The ship was assembled in space with materials originating from Earth, the Moon and mined asteroids. The *Endurance* is a native of space itself.

No easy getaway

Using a traditionally fuelled rocket to get off the Earth was both expensive and risky. There are two ways to get an object into space. You can throw it, or you can push it. In practice, we usually go for the latter, but first it's worth taking a look at the former.

If you can throw something faster than 'escape velocity', it will get away from the Earth's gravitational pull and not return. If you had a suitable superhero to help you out, they would have to throw a ball straight upwards at 11.2 kilometres (seven miles) per second for it to reach this speed. That's extremely nippy. The fastest fighter jets of the early 21st-century flew at around three times the speed of sound, but the ball would need to travel eleven times faster than this.

There is a way to cheat a little, because helpfully the Earth is rotating and we can make use of that. Something that is shot off the Earth in the right direction does not have a standing start, because it is already travelling at the speed of rotation of the Earth's surface. By piggybacking on the Earth's movement, we can get escape velocity down to around 10.8 kilometres (6.7 miles) per second – but that is still ridiculously fast. This, incidentally, was the approach taken by one of the first science fiction space flight stories, Jules Verne's *From the Earth to the Moon* (*De la Terre à la Lune*)*.

In his novel, Verne's adventurers were shot from the Earth using a 274-metre (900-foot) long cannon called Columbiad. The distinction between a cannon and a rocket is that the projectile in a cannon is only being accelerated while it is in the barrel. As soon as it leaves, it can only get slower. Unfortunately, to get a capsule up to escape velocity by the time it had traversed the Columbiad barrel would have required so much acceleration that the astronauts would have been mashed into jelly. Even if Verne had stretched Columbiad to ten kilometres (6.2 miles) in length, those on board would have suffered 600 times the force of the Earth's gravity. The acceleration they endured would be vastly more than the around 9g** that is about the most a human can survive.

Yet we've all seen video of rockets taking off and getting away from the Earth. They seem to climb very ponderously into space. Although some of the slowness is illusory, they don't fly at anywhere near escape velocity. The reason they can travel relatively slowly and still get up into space is that they aren't thrown like a projectile from a cannon. All the time that the rocket motor is active, the capsule is being pushed. And as long as the force of that push is bigger than the force of gravity pulling the spaceship down, you can travel as slowly as you like away from the Earth. It's the same as riding up in a lift. We don't need to go

---

* We need to emphasise that Verne's fictional space flight was one of the first employing science as it had plenty of predecessors that used fantastical means, such as being pulled into space by a flock of birds or using the power of evaporating dew, employed in books published in 1638 and 1657 by Bishop Francis Godwin and Cyrano de Bergerac respectively.

** The acceleration due to gravity on the surface of the Earth is labelled 1g, which is an increase in velocity of around 9.8 metres, or 32 feet, per second each second.

quickly; we just need to have enough upward force to overcome the downward pull of the Earth.

The catch, though, with rockets is that the more mass the object has, the more fuel it takes to keep it moving long enough to escape the Earth's gravity well. And every drop of fuel you have onboard adds to the mass. So that takes even more fuel. This is why rockets that carried any sizeable payload used to have multiple stages. (This was before the use of modern nuclear or antimatter-powered orbital shuttles.) That way, once the fuel in one section is mostly used up, a large chunk of the mass could be dropped off in the form of a stage, leaving far less mass for the remaining fuel to propel. Add in the fact that traditional rocket fuel is highly inflammable and potentially explosive and it can be seen that using a straightforward rocket to get into space was always a last resort.

That stages would be necessary for manned flight was publicised by Russian rocket theorist Konstantin Tsiolkovsky as early as 1903, the year that the Wright Brothers first flew an aircraft. While we're dealing with rocketry, it's also worth mentioning that to begin with, a number of respectable (if scientifically ignorant) figures doubted that rockets could work at all in space. Back in 1920, US rocket pioneer Robert Goddard published a paper entitled 'A Method of Reaching Extreme Altitudes' that suggested a rocket might be used to get to the Moon. *The New York Times* could not resist teasing Goddard in an editorial published on 13 January 1920 for what the article's writer believed was a silly error:

> That Professor Goddard, with his 'chair' in Clark College and the countenancing of the Smithsonian Institution, does not know the relation of action to reaction, and of the need to have something better than a vacuum against which to react – to say that would be absurd. Of course he only seems to lack the knowledge ladled out daily in high schools.

What *The New York Times* misunderstood when it shot itself in the foot with this article was the meaning of the 'equal and opposite reaction' bit in Newton's third law of motion (him again). The rocket does not somehow press against the atmosphere and get pushed forward in the resultant reaction. Instead, it pushes out its exhaust and the equal and opposite reaction is that the rocket is propelled forward. This can happen just as well in a vacuum as in an atmosphere. Better, in fact, as there is no air resistance to hold the rocket back.

Rockets definitely do work in space. *The New York Times* has been proved wrong many times since the 1920s (and the newspaper did issue a belated 'correction' in 1969 when Apollo 11 was on its way to the Moon). But 20th-century rockets were still clumsy, costly and dangerous. Back then, there was a lot of excitement in theoretical space travel circles about space elevators, particularly after they were used as a central feature of Arthur C. Clarke's 1979 novel *The Fountains of Paradise*. Before we get to elevators, though, we need to understand what an orbit is.

Orbital velocity

Orbits will feature regularly on our exploration of space. Understanding orbits is not particularly difficult, but they are counter-intuitive. When a spaceship, say, is in orbit, it is in freefall towards the planet or other body it is orbiting. It would be dropping straight downwards to crash on the surface were it not also moving sideways at just the right speed so that it keeps missing the planet. The outcome is that it travels around the planet at a constant height.

For any altitude above the planet's surface, there is only one speed at which the orbiting ship can travel. If it went any faster, it would fly off into space – and if it went slower, it would spiral its way down and collide with the ground.

One particularly useful orbital speed is to travel at the same rate that the planet rotates. This keeps the spaceship (or satellite, for example) at the same position over the Earth. When satellites are in this orbit around the Earth they are called geosynchronous*. Rather than moving through the sky as seen from the planet's surface, the satellites stay over the same point at all times. This can be useful, for instance, for communications purposes. To stay in place, the satellite needs to be positioned 35,786 kilometres (22,236 miles) above the Earth's surface. To put that distance into context, the circumference of the Earth is around 40,000 kilometres (24,900 miles). In fact, the kilometre was originally defined as 1/10,000th of the distance from the North Pole to the equator through Paris. So, a geosynchronous satellite has to travel upwards by nearly as far as a journey around the Earth.

Once we can imagine a satellite in a geosynchronous orbit, we have the starting point to imagine building a space elevator.

Climbing an elevator to heaven

The name sounds so simple – a space elevator is a lift that we can ride up into the sky to reach orbit. If we could set this up, there would be no need for our ascent vehicle to be laden with fuel. Let's imagine we had a nice, chunky geostationary satellite and dropped a cable from that down to the surface of the Earth. Then we simply create a vehicle to begin climbing up the cable, powered by electricity provided from that cable, so the elevator

---

* Satellites that maintain a fixed position in the sky also tend to be called geosynchronous when orbiting a planet other than Earth, but strictly the 'geo' part makes it Earth-specific. Around Mars, for example, they should be aressynchronous.

has no need to carry fuel onboard. It's a neat solution, but there are one or two problems.

Before we get into the detail, it's worth saying that the space elevator wouldn't get us entirely away from Earth. But at 35,786 km (22,236 miles) up, the pull of the Earth's gravity would be reduced to about 1/44th of the level on the surface – this means it would take very little effort to escape its pull. (Our starship also has to leave the Sun's gravity as well, requiring some more energy to be expended, but an even smaller amount.)

A first problem with the technology is that a space elevator would be an extraordinarily slow way to start a journey across the galaxy. Remember, heading up the elevator would be the equivalent of taking a journey almost around the Earth's circumference. It would feel decidedly slow, as we tend to underestimate the distance involved. If the elevator travelled at a reasonable 200 kph (124 mph), it would take about 7.5 days to make the climb.

However, the bigger issue the designer of a space elevator faces is the strength and mass of the cable. One immediate problem is that as soon as we suspend a cable from the satellite, the combined body is no longer orbiting at the right height. As Newton realised, something orbiting acts as if all its mass is in a position where there is the same amount of mass in all directions. We would need to find this centre of mass for the satellite plus cable combined – which would be below the orbital height of the original satellite. In practice, the satellite would need a big counterweight above it to counter the mass of the cable below and keep its centre of mass 35,786 kilometres (22,236 miles) above the surface.

As for the cable, its mass would be considerable. Let's assume that it is about 28 millimetres (just over an inch) across. This would enable it to carry around 50 tonnes of load, which would probably be enough for the size of elevator that we need to haul people and freight up to our starship. Unfortunately, though, 35,786 kilometres of this cable would have a mass of around 115,000 tonnes, which would mean that it would be incapable of supporting its own weight.

Even with the strongest, lightest material currently available – the atom-thick carbon-film layers called graphene – it would not be practical to make an Earth-based space elevator. And although material science has moved on since the 21st century, we still have nothing strong enough.

However, given the Moon's much lower gravity it might seem that it would be useful to build a space elevator there, enabling sections of our starship to be constructed under the lower gravity on the surface of the Moon and hauled up into space. Unfortunately, though, the chances are that such an elevator would not work on the Moon either. Although there is less gravity to give the cable weight, it would have to be longer at around 88,000 km (54,700 miles) to stay in position. To make matters worse, it could only be located on the far side of the Moon: otherwise, the relative closeness of the top of the elevator to the massive Earth would make the whole structure gravitationally unstable.

There are alternatives to rockets to get cargo off the Moon, though. The starship *Endurance* was in part assembled from sections flung up into space from the lunar surface using a mass driver. This is a device that uses electrical energy to accelerate the cargo down a long track, building up enough speed to achieve the Moon's relatively low escape velocity of 2.38 kilometres (1.48 miles) per second.

Beam me up

As we're getting our heads around the science of starships, it would be remiss not to consider the possibility of using something equivalent to a *Star Trek* transporter to get off a planet. This idea was not originally based on science, but rather on the TV show's budget. The producers couldn't afford the time and effort to produce the special effects required to show a shuttle landing and taking off every time the crew visited a planet, so opted for a magical instant fix in which people were 'beamed up'. But is this kind of transporter scientifically possible?

The only scientific avenue that might be able to help is what's known as quantum teleportation, which effectively means making an identical copy to a quantum particle in a remote location. But the reality of using this approach falls down due to the sheer scale of everyday objects when viewed at the atomic level.

Think, for example, of what would be needed to transport a human being. If you are an average size, you will have around 7×10^{27} atoms in your body. To use some form of teleportation, you would have to somehow scan every atom in your body (including its structural links) and reproduce them all at the destination. There is no clear way to do this. But even if there were, it's problematic. Let's imagine you could do this at, say, a trillion atoms a second. This sounds impressive, but it would take around 200 million years before your journey would be completed.

There is also the minor problem that transporting atom by atom doesn't mean that the finished item is then magically reassembled, whether it's you or part of a starship. And if it were you, even if the process could be made to work, and such a teleportation device could make a perfect copy at the remote location, the original person would be destroyed in the process. After passing through the transporter, you might get something indistinguishable from the original Captain Kirk, say, but from his viewpoint, each time he beams down or up, he dies. It does not sound an appealing prospect.

The conclusion is that the materials to build our starship need to be shipped up into orbit using conventional propulsion, as would any supplies and passengers, though as already mentioned, this is now more routine – far safer and less dramatic than rocket launches in the early days. All we have to do once we are pretty much out of the Earth's gravity well is to keep our passengers comfortable and to deal with the entirely non-trivial issue of travelling faster than the speed of light. But first, let's take a look at staying comfortable.

Defying gravity

We are so used to living in the vicinity of a massive object – the Earth – that it's easy to lose sight of how important gravity is to us from the point of view of both health and comfort. Before interstellar travel became possible, the most familiar trips were to the Earth's orbit, the Moon and Mars. Earth orbit might have seemed impressive when it was first achieved, but it is not a true space journey. If we look, for instance, at a famous orbiting destination of the distant past, the International Space Station (ISS), visitors onboard that makeshift craft felt that they were experiencing zero gravity – but in reality, they were in free fall.

As we've already seen, an orbit is a balancing act between falling and travelling sideways in order to keep missing. Anyone in free fall under a gravitational field does not gain any weight from that field – they feel that they are floating. But that doesn't mean that they are not experiencing gravitational attraction. The old ISS orbited at a mere 350 kilometres (218 miles), give or take, above the Earth, which means that it experienced around 90 per cent of the gravity at the planet's surface. It was only because astronauts were constantly falling that they felt weightless.

On the Moon or on Mars, by contrast, the situation is much the same as on Earth – standing on the surface, travellers feel a constant gravitational pull from the nearby massive body. As we've already seen, that is around one-sixth of the Earth level on the Moon and it's about two-fifths on Mars. Once you are on the starship, though, and well away from a planet or star, there will be no natural gravitational pull.

For a short time, this can be enjoyable. Sections of the *Endurance* are left without gravity so that passengers can experience floating around and can play three-dimensional sports. However, there are good reasons to avoid staying without gravity too long. Some find the experience induces nausea, and no one enjoys the requirement to use a bathroom with no gravity to help things progress naturally. More importantly still, the human body evolved to exist in a gravitational field of around 1g. It struggles in zero g. Muscles start to waste away, while bones lose density, making them more fragile. Long exposure to low gravity would mean that our lungs would become less effective because the diaphragm in the chest shifts up and the liver floats upwards, leaving less space for breathing.

It's not just humans (and other animals) that deteriorate under low gravity. Plants get confused and don't grow as well as usual because they use gravity to point their roots in the right direction. Way back, on one of the space shuttles originally used to access the ISS, it was discovered that a batch of fertile quails' eggs failed to hatch without gravity, which it seemed was necessary to pull the egg yolks to their proper position near the bottom of the shell to spur on the development of the chick. Amusingly, this experiment was sponsored by the then popular fried chicken fast-food company KFC.

The importance of gravity for the comfort and safety of passengers means that a ship like the *Endurance* needs to provide some form of artificial gravity. Science fiction usually provides this as a convenience at the flick of a switch with a hand-waving description of an 'artificial gravity field'.* In reality, there is only one effective approach to producing artificial gravity, which is based on Albert Einstein's equivalence principle.

This concept was the inspiration behind Einstein's general theory of relativity, which describes how massive bodies produce the effects of gravity by warping space and time. Thankfully, we don't need to get into too much depth on this theory (yet – we'll need a bit more when we encounter some of the more dramatic natural space phenomena). But the important realisation that Einstein had – something that he described as his 'happiest thought' – was that someone in free fall, for instance in a falling lift, would not feel their own weight. The acceleration they experience is equivalent to a gravitational pull and the two cancel each other out. This, as we have seen, is why astronauts are weightless in orbit.

Turning the effect on its head, if you accelerate someone, they will feel the effect of a gravitational pull. When, for example, a car or plane accelerates off at a high rate, you are pushed back into your seat. The sensation is just the same as being pulled towards the Earth by gravity – the two are indistinguishable. If you accelerate someone in space with a force of 1g, producing the acceleration they would experience due to gravity at the Earth's surface, they will feel that they have their normal weight.

In principle, then, it's easy enough to generate gravity – just keep the ship under constant acceleration of 1g. But as we will discover, conventional acceleration isn't involved in the way that a starship's drive works. And, even if the two types of motion

---

* Artificial gravity certainly seemed to work a lot better on the USS *Enterprise* than other, rather more basic safety equipment. The crew did not seem to have heard of seatbelts and were regularly thrown all over the bridge.

were compatible, it would take a vast amount of fuel to keep accelerating at 1g for weeks or months at a time.

Luckily, there is a particular kind of acceleration that does not require energy input once it has been initiated. This is the acceleration that is involved in spinning around.

You spin me round (like a record)

When we think of acceleration, it's usually a matter of getting faster (technically, it can also be getting slower) – but all acceleration means is experiencing a change in velocity. The 'v' word is not just an alternative term for speed. Speed tells you how fast something is moving, but velocity incorporates both speed and the direction of motion. Physicists call a quantity that has both size and direction a vector (as opposed to something like speed that only has size, which is called a scalar).

Because velocity is a vector, its value changes if we change direction, even if we carry on moving at the same speed. It makes sense if you think that acceleration is caused by giving something a push (applying a force in physicists' terms). That can just as much be done to change direction as it can to speed something up or slow it down.

As a result, we can generate artificial gravity by spinning something around. You may have experienced this on a fairground ride. There's an attraction that was popular in the old-style travelling fairs that is made up of an enormous vertical drum that you stand inside. The drum spins around, then the floor drops away – but when it does so, you remain pinned to the wall rather than falling. Acceleration's equivalence to gravity is generating an artificial gravitational pull towards the outside of the drum.

Something else you might have experienced while on a ride like this is a feeling of nausea. Your sense of balance is controlled by a system that involves fluid in the inner ear. This system is often confused by motion, particularly acceleration – producing the symptoms of travel sickness. Being spun around is a great way to produce nausea – something you wouldn't want to experience on our expensive interstellar tour.

Luckily, there is a way to spin without feeling bad. This is just as well – after all, the surface of the Earth at the equator moves at about 460 metres (1,500 feet) per second – over 1,600 kilometres (994 miles) per hour – yet we don't notice it. Make the rotating environment big enough and it doesn't generate that sick feeling. Obviously, the starship can't have the same diameter as the Earth – but that isn't necessary. And for that matter, we don't need the whole ship to rotate either.

The *Endurance* looks nothing like most of the spaceships of sci-fi movies. They tend to be slick and streamlined. But streamlining is only relevant to the design of a ship that is required to pass through an atmosphere. In space, there is no air resistance. The clumsy-looking *Discovery One* in the remarkably forward-looking 1960s film *2001: A Space Odyssey* was probably closer to the real thing than anything else that has been seen in the movies. That ship even had a rotating section to produce artificial gravity for exercise*, though the diameter of it was too small to have worked in reality.

---

* Fascinatingly, *2001: A Space Odyssey*, which was made in 1968 before high-quality computer graphics, had a real, full-size rotating set to provide the exercise section. The actor walked at the bottom of the wheel and the camera moved with the wheel, making it seem that the actor was sometimes upside down. The technical crew got in trouble for leaving items on the wheel, which would crash down when it started to rotate – in one case, a heavy wrench dropped from the wheel to nearly hit the visiting AI scientist Marvin Minsky. Minsky recalled how the director '[Stanley] Kubrick was livid and quite shaken and fired a stagehand on the spot'.

To get a full 1g of artificial gravity without the passengers experiencing nausea, the wheel would have to be about 500 metres in diameter. On the first interstellar liners, which used this technology, a compromise was made at 0.5g, giving sufficient artificial gravity to avoid significant muscle wastage, but making the wheel section more manageable in size at a diameter of 100 metres (330 feet). Even so, the design did give such ships a significant bulge in the mid-section. The *Endurance*, being the most modern in the fleet, has a smaller mid-section bulge, but the gravity provided there is not artificial: it is produced by neutron star technology, which makes it possible to avoid rotation altogether.

Speculation alert

As we take a trip on the *Endurance*, you will encounter occasional 'Speculation alerts' like this one. With the exception of the drive system, which is purely fictional, these highlight when you see or experience something that is possible based on a 21st-century understanding of science, but that may not have proved actually true or feasible by the time the *Endurance* was launched.

Unlike on earlier ships, the *Endurance*'s gravity section is not rotated. Instead, it is a large cylinder with a long, extremely thin core of neutron star material threaded through its centre. (We will be visiting neutron stars later in the voyage.)

This gives the different cylindrical floors positioned around that core a gravitational pull equivalent to between 1 and 0.2g, depending on the level's distance from the centre. Extracting neutron star material was an interesting challenge to say the least – but given interstellar flight technology, it is not scientifically impossible.

Tea, Earl Grey, hot

Few can think about starship design without reference to that 20th-century exemplar, *Star Trek*'s USS *Enterprise*. In the 'Next Generation' incarnation of the show, much was made of the replicator, which was used to produce food and drink, including the preference of Patrick Stewart's Captain Jean-Luc Picard for 'tea, Earl Grey, hot'.

The 'hot' part of this command caused some controversy, as no one would sensibly drink Earl Grey tea cold. It seems hard to believe that a ship featuring technology that could produce food and drink apparently out of thin air would not have a sufficiently clever AI to realise that when someone asked for Earl Grey tea, they wanted it hot. But what is often missed in criticising the vison of the *Star Trek* creators is how much the replicators were under-utilised. Why could they make food and drink but not, say, uniforms or weapons or parts for engineering?

There certainly is an advantage in a starship being able to carry raw materials and construct most consumables on board, rather than having to carry a huge range of stores. Unlike an ocean liner, a starship can't rely on supplies being available at every port. In reality, there are very few locations taken in by a trip like that undertaken by the *Endurance* where there will be any outposts of humanity from which to pick up essentials.

The practical precursor to the replicator, first introduced in the 20th century, was the 3D printer. Initially limited to plastics, during the 21st century, their capabilities were expanded to include everything from food to metals, in applications that ranged from tiny mechanical parts to full-size buildings. The 3D printer is still a staple for producing objects on the *Endurance*, but to get to a *Star Trek*-style replicator required inspiration from a concept put forward by the American engineer K. Eric Drexler.

It seems that Drexler was himself inspired by the great American physicist Richard Feynman. Although Feynman was best known for his work on quantum physics and on the commission that investigated the disaster that occurred in the flight of the Space Shuttle Challenger, he also explored the science of manipulation of the very small in a 1959 lecture 'There's Plenty of Room at the Bottom'.

Feynman envisaged using extremely small manipulators to construct even smaller manipulators, which then worked on the next level and so on, until devices existed that could interact with individual atoms to construct anything from the appropriate chemical elements. This wasn't intended as a practical piece of technology – Feynman was aware of the physical challenges involved – but he put it forward as a provocation to think differently. This was evidenced in a couple of challenges that Feynman threw out to the audience: to build a motor smaller than 400 micrometres (0.016 inches) on each side and to produce text small enough to fit the entire *Encyclopaedia Britannica* on a pinhead.

Remarkably, the first of these challenges was achieved the next year (although using conventional engineering methods), while the second was cracked in 1985. But neither went to the extreme that Feynman envisaged. In his 1986 book *Engines of Creation*, Drexler came up with a more detailed description of how nanoscale manipulation of materials could be undertaken with a collection of tiny robots called assemblers that could piece together materials atom by atom.

Just as the *Star Trek* transporter has a scale problem, so did Drexler's vision. There would have to be vast numbers of assemblers to do anything realistic. Imagine you had sufficient assemblers to put together, say, 7,000 trillion atoms per second – it would still take 200 years to construct Picard's cup of tea.

Speculation alert

It is entirely possible that the kind of assemblers envisaged by Drexler, using an engineering approach, will never be physically possible to construct. The forces and interactions between objects on the scale of atoms are totally different to those we experience in the everyday world. However, nature makes use of 'molecular machinery', for example, in manipulating molecules in the production of proteins using the templates provided by genes. Assemblers are likely to be possible at some point to create pseudo-organic materials, but may not be able to produce, say, a piece of electronics. Biological processes can also take a considerable amount of time to complete an assembly (think how long it takes to produce a fully grown human), so such biological assemblers are likely to be limited to small products when working on a short timescale.

On the *Endurance*, a mix of onboard stock and 3D printing is complemented by the use of biological assemblers for small quantities of essential substances, such as medication and the flavouring used to give a relatively convincing facsimile of familiar foodstuffs. As yet, no one has ever requested Earl Grey tea, hot or otherwise.

The light speed barrier

It's not enough to survive and thrive in space – a starship, by definition, must be able to reach the stars. And it is in the physics of high-speed motion that we hit the biggest physics problem for interstellar travel: relativity. Einstein's special theory of relativity makes it clear that there are some limitations of movement through time and space, all tied into the limiting speed of light in a vacuum.

At the start of the 20th century, it had already been established that in order for light to exist, it has to travel at a fixed speed in any particular medium*. This makes it different from anything else. If two cars travel towards each other, both moving at 50 kph (31 mph) we add their speeds together and say their relative speed is 100 kph (62 mph). But however fast you move towards or away from a beam of light, it will always travel at the same velocity. In his special theory of relativity, Einstein established an implication of this fact when combined with basic physics: if something moves with respect to an outside observer, that observer will see the object's time slow down, its length shrink and its mass increase.

It may seem extremely unlikely to be able to get to this finding from the simple fact that light always travels at the same speed, but the maths involved requires little more than Pythagoras' theorem and can be followed easily by a high-school maths student. We won't go through the details right now, but they are included in an appendix at the end of the book for those who are interested. We will also come back to relativity and its implications for time travel towards the end of our trip.

The limitations of relativity mean that in practice it is very difficult to get a spaceship up to anything near the speed of light. And it is physically impossible to accelerate past the speed of light, which presents starship designers with a problem. Let's say we manage to get a ship up to a respectable quarter of the speed of light. From the viewpoint of those on Earth, it would still take around sixteen years to reach the nearest star to Earth

---

* What is usually described as the speed of light (around 300,000 kilometres or 186,000 miles per second) is actually its speed in a vacuum, which is its fastest rate of travel. When light is moving through air, or water, or glass, for example, it has a lower (though still very fast) speed.

other than Sun (Proxima Centauri) and another sixteen years to return.

There would be some slowing of time due to special relativity – but from the viewpoint of the travellers, at this speed, the round-trip journey would involve spending over twenty years onboard the ship. Even if the starship could travel so close to light speed that time hardly passed at all while travelling, a full eight plus years would have elapsed on Earth when they returned, meaning all their friends and relations would have aged in comparison to the travellers.

As it happens, Proxima Centauri will be on our itinerary a little later, but to get the true Interstellar Tours experience we need to travel much greater distances, which means being able to fly much faster. Sub-light speeds would not hack it. But there is a potential solution – because, contradictory though it sounds, although it is not possible to pass through space at greater than the speed of light, it is entirely possible (in theory) to get from A to B at much greater than light speed.

Let's do the warp drive again

Even today, a whole lot of scientists are *Star Trek* fans, so it was perhaps no surprise that some have dreamed up equivalents to the *Enterprise*'s warp drive based on actual physics. We know from the way that the universe has been able to expand at much faster than the speed of light that space itself can move faster than light – the light speed barrier only applies to movement *through* space. What was needed was a way to travel by transforming the space around a ship.

Of course, warp drives as shown in science fiction don't need to work in the real world, and certainly nothing was going to happen by using dilithium, the fictional material that makes *Star Trek*'s warp drives work. (There is such a thing as dilithium, but it's nowhere near as exciting as the *Star Trek* version.) But in the mid-1990s, the Mexican physicist Miguel Alcubierre, working at the University of Wales in Cardiff, suggested a way to propel a spaceship by warping the spacetime around it.

The so-called Alcubierre drive uses a hypothetical approach to squash up spacetime in front of a starship and stretch it behind the ship. Imagine the ship is an ant trying to get from one end of a strip of rubber to the other. It could take the long route, simply walking along the rubber strip. But with some outside help, the rubber in front of it could be squashed inwards and the rubber behind the ant stretched out. The 'rubber band space' would still be the same length, but now the ant, without taking a step, could be much closer to the other end.

In the hypothetical equivalent of Alcubierre's warp drive, the ship does not strictly move through space at all, but instead its position in spacetime changes as a result of the manipulation of local space performed by the drive. The original paper was more theory than anything practical, but by 2012, an American NASA physicist called Harold White, who enjoyed speculative science, had suggested that it might actually be possible to get around some of the obstacles to making such a drive a reality.

Perhaps the biggest single problem was having enough energy to make it work. Remember this is not just a matter of moving through space, but one of manipulating spacetime itself. This is something that usually requires the gravitational field of a massive body like a star (and even then, it is a relatively small effect). A first, back-of-the-envelope calculation suggested that it would be necessary to expend as much energy as represented by the entire mass of Jupiter to perform this action, but White suggested a mechanism which would reduce the energy demand for a working warp drive to the energy equivalent of consuming around 725 kilograms (1,600 pounds) of matter. Bear in mind that the equivalent energy of matter is calculated using $E = mc^2$, where c, the speed of light, is a huge number.

Even this amount of energy is almost unimaginably vast – but there were bigger issues to combat too. The drive's design depends on using a concept known as negative energy, something that does exist, but is usually only available in extremely small quantities. And Alcubierre has pointed out that the ship would have to remain out of contact with the exterior of the warp bubble around it, as otherwise it would have to be able to communicate at faster than light speeds – which isn't possible – so it would be flying blind.

Even if the drive itself proved possible to build, the crew of a ship travelling with this kind of warp drive would find it impossible not to crash into things. An Alcubierre drive ship would not disappear from one location with a pop and appear magically in another, as happens in all the best science fiction movies. It would be warping its way through ordinary spacetime. This would mean that anything in the way – stars, planets, asteroids or even a few specks of dust – would produce catastrophic collisions and impressive explosions. Wonderful though the wishful thinking is, this kind of warp drive seems highly unlikely ever to work.

There are similar problems with another science fiction favourite to get around relativity: using wormholes as a means of long-distance travel. A wormhole, also known as an Einstein–Rosen Bridge, is a hypothetical structure where a portion of spacetime gets so distorted that two points in space that can be many light years apart are connected as if they were very close together.

Imagine our illustrative ant is now travelling from one end of a piece of paper to another. To make an analogy to a wormhole, we would fold the paper over so that the end and the beginning are near each other. Then we would make a breach through the paper near the end and the beginning and join those holes together so the ant could pass through the hole and get to the other end without travelling through the space in between.

We can conceive of half a wormhole in the form of a black hole that can form when a star collapses (we'll visit one later), which is such an intense distortion in space and time that it can be envisaged as leading to another location – but the problem is that a black hole is a one-way gateway. You can get in, but you can't get out. To get out of the other end you would need a conceptual entity called a white hole, which is kind of an anti-black hole you can get out of but not enter, and then somehow you would have to link the two distortions of spacetime together. But no one has ever seen a white hole.

Although a wormhole is theoretically possible, none have ever been observed in the natural world and it is hard to conceive of a practical way to construct one. To make matters worse, the physics tells us that without significant amounts of negative energy (just as unlikely here as for a warp drive), a wormhole would collapse the moment that anything tried to go through it. Wormholes also appear to be a dead end for faster-than-light space travel.

Space that's gone hyper

The *Endurance* uses a mechanism that was not devised until the 22nd century, based on a whole new view of quantum gravity. Back in the 20th century, it was realised that of the four fundamental forces of nature, one was very different. Three of the forces – electromagnetism and the strong and weak nuclear forces – are quantised. This means that they are transmitted by 'carrier' particles, effectively coming in tiny chunks. In electromagnetism's case, these particles are photons. These forces are compatible with quantum theory, arguably the most successful scientific theory in history.

However, gravity is different. The 20th-century theory of gravity, Einstein's general theory of relativity (which we will discover more about later), is not quantised. In effect, according to the general theory, gravity isn't a force in the same sense as the others are. According to the general theory, anything with mass warps the space and time nearby. The more concentrated the mass, the bigger the warp. And it's these twists in space and time that make massive things move – giving the effect of a force.

These two theories – the general theory of relativity and quantum physics – are incompatible. But physicists hoped for many years to find a unified theory that combined all of the fundamental forces. By the early 21st century, there were two prime contenders, string theory and loop quantum gravity. But neither seemed entirely likely to prove correct. The field was ripe for a totally different approach to viewing gravity and quantum physics together. The *Endurance*'s faster-than-light drive is only possible because the resultant theory made it possible to interact with hyperspace.

> **Science fiction alert**
>
> As yet, there is no science-based solution to the faster-than-light travel problem. All the best scientists have a 'never say never' attitude – but it is not clear how such a solution could be reached. It ought to be stressed that, as of 2023, there is no evidence for the existence of hyperspace: it is a science fiction concept, and the *Endurance*'s hyperspace drive is a piece of science fiction technology.

Hyperspace is the third mechanism typically used in science fiction to get around the light speed barrier, along with warp drives and wormholes. The idea is that there exists a 'non-space' that somehow maps differently onto locations in real space. This means that by taking an appropriate shortcut route through hyperspace, it is possible to get from A to B without travelling far at all. The *Endurance*'s drive is limited to distances on the scale of our galaxy, but the time taken for a jump is not related to the distance involved. All trips through hyperspace take 42 minutes. No one knows why[*].

It's good to talk

One last consideration from faster-than-light travel is the impact it might have on speed of communications. The hyperspace drive is not about transmitting something ultrafast, but rather involves getting from A to B using a mechanism that circumvents the need to travel faster than light though normal space. Such a technology would not enable electromagnetic signals to be sent from A to B faster than light.

As a result of this limitation, there is no 'hyper radio' connection to the *Endurance* from Earth. The only way to get a message between the ship and our home planet is to carry it physically on another ship or an unmanned probe. In this sense, the voyages of the *Endurance* are more like the experience of pre-radio ocean travel on Earth than they are of a 21st-century plane or spaceship. But rather more luxurious.

---

[*] Fans of *The Hitch-Hiker's Guide to the Galaxy* might be pleased to know that 42 is indeed the answer to the ultimate question of life, the universe and everything when it comes to taking interstellar voyages.

Departure time

Passengers on the *Endurance* will encounter a wide range of scientific wonders as they voyage across the universe. Of these, a surprising amount was already known by the 2020s, despite the immense distances involved and even though very little had yet to be directly explored, whether by manned or unmanned probes.

Before the *Endurance* takes us into interstellar space, though, we will come across a few reminders of earlier human attempts to get out into space. By making short hyperspace jumps, our starship will be able to catch up with relic probes, still making their slow but steady way into the depths of space. They will give us an insight into humanity's earliest endeavours to reach for the stars.

PASSING THE COMPETITION 2

Taking our first short jumps through the solar system, there is a chance to spot some of our predecessors in space exploration as we catch up and then overtake them, leapfrogging them on their way out into the wider galaxy.

As you settle into your seat in the main viewing lounge of the ship, the lights dim, and the walls become transparent. Don't be surprised if there is a gasp from your fellow passengers, as you now seem to be floating in space. It is normal to initially feel a little queasy at this point in the experience. If you find it too unsettling, please close your eyes for a few moments. If the view is still a problem, touch the seat controls and you will be guided out of the lounge – but guests are recommended to take time to acclimatise, as this view will be used throughout the voyage.

There is something very special about the way that the viewing wall works. Ever since the introduction of film and TV, we have become used to virtual views – appearing to be looking at something, when in reality, we are looking at a screen featuring light generated from photographic images or electronic signals. In video, for example, light from the scene we are viewing is converted by a camera into electrical signals. These are transmitted, usually using electromagnetic waves, to a receiver where the signal is then converted back into light. However good the camera and the screen, this gives a feeling of being distanced from the view. However, the *Endurance*'s viewing lounge uses the process known as quantum teleportation to produce the images that you see.

Here, incoming photons of light from outside the ship are captured by detectors in the vessel's outer panels that then generate identical photons inside the walls that fly inwards. These new photons are identical at a quantum level to the incoming photons. We don't see a reproduction of what is outside. We see the light from outside as if it were able to pass straight through the ship's wall without the metal hull being there. Of course, the system is capable of more sophisticated action than mere transparency – the viewing walls can zoom in, change light from outside of the visible spectrum, such as radio or X-rays, into a visible image or simulate a view that isn't really there – but when the walls appear to be transparent, they are truly so in a far more visceral way than is the case with a camera and screen.

Even when light passes through a sheet of glass, the process involves absorption and re-emission of photons, degrading the original. The viewing walls on the *Endurance* give a more direct image of the outside than would a transparent window, without the safety implications of having to make holes in the hull.

The pioneer

All the probes we are going to encounter as we head out on our journey have now left the solar system and are technically already in interstellar space. Exactly where the border is remains an arbitrary location. There is no sign saying, 'You are now leaving the Solar System' to mark our transition. One common measure of the limit of the solar system is the extent to which the solar wind – the flow of particles constantly emitted by the

Sun other than light – sweeps away the interstellar medium, which is the gas and dust usually found in between the stars.

Known as the heliosphere, the nearest edge of this region to Earth is about three times further out from the Sun than Pluto – around 90 astronomical units (AUs) where 1 AU is the distance from the Earth to the Sun. Despite the name, the heliosphere isn't really spherical, but rather is shaped like an elongated teardrop*. Even so, we can consider the limits of the solar system to be located around that 90 AU mark.

Drifting alongside us now is one of the first two missions ever to explore the outer solar system – NASA's Pioneer 11. The most prominent part of the probe is its communication dish, but you can also make out a range of components, including a magnetometer, high above the craft on a boom, the red asteroid and meteoroid detector and the pair of nuclear electric generators in their finned containers. The whole thing weighs in at just 259 kilograms (570 pounds) and was launched on 5 April 1973.

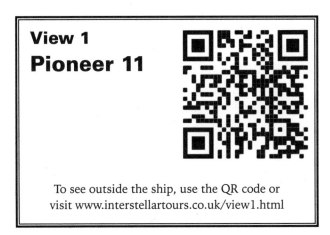

View 1
Pioneer 11

To see outside the ship, use the QR code or
visit www.interstellartours.co.uk/view1.html

---

* Even if it weren't for the name, you might expect the heliosphere to be spherical, but the solar system is moving through the interstellar medium as the Sun orbits the galactic centre. As a result, the heliosphere is naturally streamlined: squashed in front of the Sun's direction of travel and extended behind it.

> ### A quick note on the QR codes
> We have provided high-quality colour photographs on the website accompanying *Interstellar Tours* as souvenirs of many of the views you will experience on the voyage. Point your phone or tablet's camera at the code to be taken to the relevant view. Alternatively, type in the link to see the image.

You might think that the first of the probes we meet, the one that's closest to Earth, would be the last to have been launched, but Pioneer 11 has proved to be one of the slowest of our long-range early missions. It has been out of contact with the Earth since 1995 – but thanks to the lack of reactive material in this part of space, it looks pretty much as it did when it was first launched back in 1973. Pioneer 11's primary targets were Jupiter and Saturn, including encounters with a number of these planets' moons.

Pioneer 11 was part of a series of NASA Pioneer probes launched between 1958 and 1978. The first few were (mostly failed) lunar probes. Between 1965 and 1968, Pioneers 6 to 9 were used to study space weather, particularly solar wind. Pioneer 10 was pretty much a twin of Pioneer 11 that headed to Jupiter and beyond, while a couple of Pioneers from later in the 1970s were sent to Venus.

Along with the slightly older Pioneer 10, the probe we can see features an early attempt by NASA to give any alien life that might encounter it a hint to Pioneer's makers. If we zoom in on the area around the antenna support struts, you can see a metal plaque, partly concealed as it is positioned to be protected from the impact of dust. The plaque is made of gold-plated aluminium and is about 23 centimetres by fifteen centimetres. Here it is in all its glory.

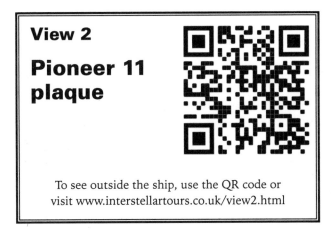

View 2

Pioneer 11 plaque

To see outside the ship, use the QR code or visit www.interstellartours.co.uk/view2.html

Talking to aliens

Any attempt to communicate with a non-human audience involves an impressively large degree of guesswork. As we will discover, it also requires aliens to be experts at solving puzzles that might tax the most intelligent human.

Along the bottom of the plaque, you can see a symbolic layout of the Sun and the planets of the solar system, including Pluto, which was downgraded from being a planet in 2006. An arrow details the Pioneer probe's flyby of Jupiter and Saturn, though unfortunately, this is only relatively accurate for Pioneer 10, as Pioneer 11 took a different trajectory. This part of the plaque seems fairly straightforward, though even this requires a particular human approach to the visual representation of objects and motion in the form of the arrow. However, apart from this and the images of naked humans (which caused some concern in the relatively puritanical US culture that existed at the time of the launch), it can be a very significant challenge even to other humans to work out exactly what else is going on here.

Imagine how much harder it would be for an alien without human terms of reference. Strictly, there is one other straightforward image – behind the people is an outline of the shape of the Pioneer probes, which is to scale with the humans. However, assumptions are already being made here. Because of the way we recognise shapes, the probe diagram stands out fairly well to us. But to aliens with different optical wiring, it could simply jumble up the more complex imagery that is drawn across it, confusing the picture of humans.

Up to this point, we have looked at the plaque's representations of visually accessible things like people and planets. But in the top left of the plate, we are supposed to be able to deduce that what we are seeing is a diagram of a hydrogen atom, which is undergoing a particular type of change known as a hyperfine transition where its electron spin is flipped. I suspect that if asked what this image is supposed to represent, this would not come anywhere near the top of the list of possibilities for most people. The two circles represent hydrogen atoms, with the line at the top, ending in a dot indicating the spin direction. Not especially clear.

The spin-flipping atom is intended to provide a useful key to the rest of the diagram – the line between the two versions of the atom represents both the length of 21 centimetres (8.3 inches) – which is the wavelength of light that has the atom's spin-flip frequency – and the time duration of 0.7 nanoseconds (billionths of a second). This is how long it takes for that flip to occur, but it seems highly unlikely that our putative aliens could ever deduce this.

Given this very detailed clue as to what is going on, why not try to deduce the significance of the markings down the far right-hand side of the plaque. Any thoughts?

The size key from the hydrogen atom is supposed to be applied to work out the size of the female human, hence the two small horizontal lines aligned with her head and feet. The number of 21-centimetre units required is identified by the three vertical dashes* and one horizontal dash to the right of her midpoint. The resultant inscription of ---| is intended to be read as 1000 – which is binary for eight, indicating that her height is 8 x 21 = 168 centimetres (66 inches). But you would have to be extremely lucky to guess that those dashes represent 1000.

The final detail of the communication comes in the pretty but obscure-looking set of lines originating from a point on the left-hand side. The lengths of fourteen of these lines indicate distances from Earth to various pulsars, with the binary numbers on each line indicating the period of each pulsar in multiples of 0.7 nanoseconds. (Unfortunately, one of these shows an incorrect value.) The last line, heading off behind the people, is supposed to indicate our solar system's position with respect to the centre of the galaxy. Anyone who spotted all of this without help is entitled to collect a gold-plated aluminium 'I'm a genius' pin from the purser's office on deck 23.

The plaque was later criticised for sexism because it is the male human who seems to be taking charge of the image by doing the waving, although Carl Sagan, who thought up the detail of the plaque with astronomer Frank Drake, and the plaque's designer, Sagan's first wife Linda Salzman Sagan, certainly never intended this.

---

* Note that the actual plaque shows three identical vertical dashes – some reproductions of it have accidentally split the lowest of the three into two parts. As we have had to add to the direct imaging because the plaque is partly concealed by the structure of the probe, this error is reproduced here.

Discovering New Horizons

Soon after leaving Pioneer behind us, we come across New Horizons. As the different style of name suggests, this is a significantly later probe, launched in 2006. There's more technology onboard than there was on Pioneer 11, and it has nearly twice the mass at 478 kilograms (1,050 pounds), although the general appearance of the probe is similar. Like its predecessor, this craft undertook a Jupiter flyby, sending back more detailed images, but its main claim to fame was providing astronomers on Earth with the first close-ups of Pluto, which, by the time of New Horizons' launch, had already had its controversial downgrade from the ninth planet to one of a range of dwarf planets. New Horizons then went on to study a number of the objects in the Kuiper Belt, a second asteroid belt that ranges from 30 to 50 AUs out from the Sun. It was designed to send back information to Earth through to the 2030s.

View 3

New Horizons

To see outside the ship, use the QR code or visit www.interstellartours.co.uk/view3.html

It was hoped before its launch that New Horizons would carry a much more comprehensive message to the wider galaxy than Pioneer 10 and 11. By 2006, information technology was far more advanced, and the probe was supposed to carry a 150-megabyte message in digital form, but in practice this didn't happen. Storage of 150 megabytes might seem practically non-existent: even as early as the 2020s, it wasn't unusual for a phone to have 500 times this amount of capacity; but it was a large amount when New Horizons was conceived. The intention was that the digital content would be collected from many different countries and cultures around the world. Instead, the probe carried the space-travelling equivalent of a high-school time capsule.

The randomly attached bits and pieces included a CD-ROM containing over 430,000 names (more as a publicity stunt than to act as an alien phone book), a pair of US stamps celebrating the New Horizons mission, two US coins (representing the states of Florida and Maryland where the spacecraft was built) and a nylon US flag. To add to the impression that the probe carried the contents of someone's attic, there was also a small piece of SpaceShipOne, the suborbital launcher that would become part of Virgin's attempts to send tourists near to the edge of space – a venture that bears as much resemblance to the voyage of the *Endurance* as a single blade of grass does to a rain forest.

New Horizons also carried a small container with the following label on it:

Interned herein are remains of American Clyde W. Tombaugh, discoverer of Pluto and the Solar System's 'third zone'. Adelle and Muron's boy, Patricia's husband, Annette and Alden's father, astronomer, teacher, punster and friend: Clyde W. Tombaugh (1906–1997).

It was an impressive send-off for the discoverer of Pluto – and perhaps having simple relics of our civilisation is, in a way, more informative for aliens than obscure clues engraved on a plaque.

Even so, the range of objects included could have been a far more imaginative and less random collection of artefacts better designed to reflect humanity.

The gold standard

In our current micro-jump through hyperspace, we are passing by Pioneer 10, but we won't stop beside it as there is little new to see after visiting Pioneer 11. We're also zipping past Voyager 2, before we stop to take a closer look at Voyager 1. Launched in 1977, Voyager 2 took in Jupiter, Saturn, Uranus and Neptune and was, slightly oddly, launched sixteen days before Voyager 1. The first Voyager had a mission that included Jupiter, Saturn and the biggest of Saturn's moons, the aptly named Titan.

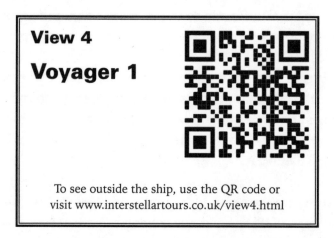

View 4

Voyager 1

To see outside the ship, use the QR code or visit www.interstellartours.co.uk/view4.html

The reason we've singled out Voyager 1 from the pair of probes is that it was the faster of the two, and so ended up being the first human-constructed object to leave the solar system. For a long time, it was also humanity's best time machine. As we've seen, due to special relativity, time slows down on any moving object as seen from Earth – the longer and faster it travels, the further it moves into the Earth's future. Although Voyager 1 has since been overtaken in this regard many times over, it was a notable front runner by having travelled 1.1 seconds into the future by about 40 years into its voyage in the early 2020s. This can be difficult to get your head around. Imagine that the time on a probe has run sufficiently slowly that time on it was a year behind Earth time, so the date on the probe was, say 1 January 2050 when the date on Earth was 1 January 2051. If the probe came back to Earth, it would find it had moved one year in the future. This is a real effect that has been experimentally proved many times.

The Voyager probe weighed in at 825 kilograms (1,820 pounds), making Voyager 1 the biggest craft we are going to see (apart from our own) on this trip, and remarkably, the Voyagers continued to operate into the mid-2020s. As we manoeuvre around the probe, you can see a bright circular object around 30 centimetres (one foot) across. Let's zoom in and take a look at Voyager's successor to the Pioneer plaque – in this case, a gold disc.

View 5

Gold disc cover

To see outside the ship, use the QR code or visit www.interstellartours.co.uk/view5.html

Each of the two Voyager probes carried one of these discs of gold-plated copper, housed in a gold-plated aluminium cover, which is what you can now see. This outer shell has some inscriptions that were carried over from Pioneer. You will have spotted the inscrutable hydrogen spin flip scale marker and the Earth location pulsar map at the bottom (telling us, apart from anything else, that Sagan and Drake were leading NASA's design team once again), but there is a whole collection of extra mysterious markings to interpret here. Any thoughts on what these markings are supposed to mean?

The top-left part of the diagram provides instructions for playing the record inside the case, which was a physically inscribed disc using the same kind of technology employed when making vinyl records. The circle is supposed to show the playing speed of the disc (one rotation lasting 3.6 seconds), using that hydrogen flip notation. Because the time for the flip is short – just 0.7 nanoseconds – this makes for a very long binary number in the | and – notation we met at Pioneer 11, though confusingly, it appears to be written in the reverse order to which data on vinyl records used to be read.

Immediately below is a view of the record sideways on, with another binary number, this time broken up into a number of segments, which supposedly specifies the playing time as lasting about an hour. (The cartridge to play the record was included with the satellite, hence the surprisingly detailed view of it in these two images). All the detail on the top right was intended to give an explanation of how to decode the video signal that was included on the record, which, if done properly, would first show a circle. This is particularly impenetrable unless you know what it's supposed to indicate.

The record itself contains a whole host of material (if only an alien could work out how to play it). There are 115 images, human sounds (which include messages from the UN Secretary General Kurt Waldheim and the US President Jimmy Carter), music ranging from Bach and Beethoven to Chuck Berry and Azerbaijani folk music, whale song and (somewhat bizarrely) an hour's recording of the brainwaves of Sagan's second wife Ann Druyan. As you can see, there are no naked people on display on the disc's case. After the complaints about the Pioneer plaque, it was decided not to include an image of naked humans, although there is a drawing among the record's contents where, this time, it is the woman who is waving.

In the rambling 1979 Star Trek movie *Star Trek: The Motion Picture*, a probe that had become self-aware, calling itself 'V'Ger', was featured. This is sometimes confused with the real Voyager missions – but the film version was a fictional, larger probe called Voyager 6, that had supposedly been upgraded to sentience by aliens. Science fiction has also occasionally dabbled with aliens finding the Voyager gold discs, though rarely with any great significance being applied to this occurrence.

Some scientists, notably Stephen Hawking, expressed concern in the early space-going period about our revealing ourselves and our location to a potentially threatening alien civilisation by sending out these physical messages. But in reality, the chances of these discs being found, decoded and inspiring aggressive invaders to find us are tiny. We can only locate them because we were in contact with them for so long and can extrapolate their current locations. Not only are they the equivalent of a message in a bottle in an unimaginably vast cosmic ocean, but they are not even the easiest way for aliens to spot human existence. Our radio and TV signals are travelling away from Earth in all directions much faster than these probes and are far more likely to be detected – although they will be extremely weak by the time they reach any potential distant civilisation.

Breakthrough Starshot

We are going to take in one last probe on a whole new scale before we move away from the human sphere of influence. Up to now, we have seen larger and larger devices, but this final one was based on a radically different concept, dreamed up in the early 21st century, called Breakthrough Starshot. Looking out from *Endurance*, it's taking us a few moments to find it, both because it is moving far faster than the other probes we have seen and also because it is so tiny.

Breakthrough Starshot was a programme launched in the 2030s, which propelled around 1,000 tiny probes up to around 20 per cent of the speed of light. This was done by building a battery of ground-based lasers, which accelerated the probes using light sails.

View 5a

Light Sail

To see outside the ship, use the QR code or
visit www.interstellartours.co.uk/view5a.html

Such sails are made from a reflective material spread out from the probe that make use of the fact that, though light has no mass, it can produce a tiny amount of pressure, and so can accelerate a lightweight object.

You can see the filmy remains of the sail around the probe still. As far as we are aware, just one of the fleet of probes did successfully make it to the nearest star system to the Earth, reaching Alpha Centauri in the mid-2050s. This is a triple-star system, the nearest of which to Earth is Proxima Centauri.

The successful probe made the first close pass of a planet orbiting another star, Proxima Centauri b, which we will see later on in our journey. The planet is now considered interesting, but at the time, it proved something of a disappointment as it lacks a significant atmosphere and shows no signs of life.

Speculation alert

In 2023, it was not known if the Breakthrough Starshot programme, founded in 2016 by a group featuring big names such as Stephen Hawking and Mark Zuckerberg, would go ahead – or, if it did, whether any of the probes that were launched would reach Proxima Centauri. It was already known that the star had at least three planets, one of which, called Proxima Centauri b, had similar proportions to the Earth. In principle, this planet could have been habitable, though there was no evidence that it had an atmosphere as of 2023, and its star is one that is likely to strip atmospheres away from a planet that initially had one.

The Breakthrough Starshot programme was apparently based on the 'streetlight search' principle. Imagine you have lost a valuable item on a very dark street. You think that you lost it at one end of the street where there are no streetlights. At the other end of the street is a single bright light. The search principle suggests it's better to look where you can actually see something than somewhere you can't, even if the probability of finding what you are looking for is lower. At the time that the Breakthrough Starshot programme was started, there were already plenty of 'extrasolar' planets orbiting other stars with a better chance of hosting life. But Proxima Centauri b was the only one located at a practically accessible distance – it was under the single streetlight.

Although the Starshot probes, known as StarChips, were only a few centimetres across, each carried five cameras, low-energy photon thrusters and a laser communicator that it was hoped would be capable of relaying data back to Earth from four light years away. As expected, the vast majority of the probes failed, either due to collisions with dust or, as the case with our probe here, losing contact with the laser before it could get up to full speed, leaving it lagging far behind most of its cohort.

Despite this, the StarChip we can see is travelling far faster than any of the other probes we have encountered and is expected to eventually make it to the vicinity of the Alpha Centauri system.

A selected sample

The probes we have stopped to take a look at are not the only ones to have left the solar system – a number of others were sent out throughout the latter parts of the 21st and into the 22nd century, but they all had the same limitations. Unlike the *Endurance*, they are not travelling under power. As a result, until they come under the gravitational influence of another star, they will be very gently decelerating because the gravitational pull of the Sun continues to have some influence on them. Each of them is travelling at greater than escape velocity for the solar system, though, and so will not return.

Unlike the *Endurance*, these probes will never have a chance to explore the wider sights and marvels that the galaxy has to offer. But now, with our local sightseeing complete, the *Endurance* can enter hyperspace for a longer distance jump and head out to the first major destination on our itinerary – the Orion Nebula.

STELLAR NURSERY 3

We have entered hyperspace and are part way through our 42-minute cruise to our first truly interstellar destination, the Orion Nebula. Before we arrive, we will see on the viewing wall how the Orion constellation looks from Earth – and within it, we will discover the Orion Nebula, a remarkable feature that was first clearly identified in 1610.

Mr Herschel's inspiration

From the Earth, a nebula appears as a small fuzzy patch of light in the sky, looking rather like a star that has been trodden on and squished. Where the Milky Way was named for its likeness to a spillage of milk in the sky, in Latin, a *nebula* is a fog or mist. For a long time, nebulae were thought to be stars that were in the process of forming (and we now know that some of them are) – but back in the 18th century, amateur-turned-professional astronomer William Herschel had different and more dramatic ideas.

As a scientist, Friedrich Wilhelm Herschel was one of a kind. He was born the son of a bandmaster in the Hanoverian Guard in 1738. As part of a brief teenage military service (as a bandsman), Herschel was sent to England with many of his contemporaries when King of England and Hanover George II suspected the possibility of a French invasion and wanted a strong military presence. Herschel appeared to have enjoyed his time in England, and after the army returned home, he left his post, moving back to Britain where he became organist to the Octagon Chapel in the then highly fashionable city of Bath.

In his spare time, Herschel developed an interest in astronomy. Helped by his sister Caroline and brother Alexander, he constructed an eight-inch reflecting telescope*, and eventually discovered what he initially thought was a comet but turned out to be a previously undetected planet. Herschel, something of a social climber, attempted to get the planet called George after the king (he was hoping to call it Georgium Sidus – George's star, which doesn't have a great ring to it). The planet wasn't formally labelled as Uranus, a name suggested by German astronomer Johann Bode, until 1860.

Astronomy overtook Herschel's life and, along with his sister Caroline, he put in many hours sweeping the skies. King George gave him the title of King's Astronomer along with financing, in exchange for Herschel moving closer to Windsor for George's convenience. Herschel settled in the nearby, then small, town of Slough and built a number of large telescopes, culminating

---

* This wasn't an easy process in Herschel's day as he had to make his own mirrors by pouring molten metal into a mould that he made from horse manure. While producing one of his biggest mirrors, three-feet (90-centimetres) across, the mould split in the cellar of his house sending molten metal onto the stone floor. The flooring split under the heat, sending deadly shards flying across the room like shrapnel: luckily no one was injured.

in his mighty 49-inch (125-centimetre) reflector, which had a 40-foot (twelve-metre) long tube, supported by a complex wooden framework of poles and ladders.

Herschel suspected that the Milky Way was a vast collection of stars, roughly disc-shaped, of which our solar system was a tiny part, just one star within that massive cluster. This picture of the Milky Way gradually became accepted as the nature of the universe – it was something much bigger than the solar system, containing millions of stars. But Herschel had an even more remarkable idea.

With Caroline's assistance, he had made a catalogue of nebulae, and in a paper, entitled 'On the Construction of the Heavens', he suggested that the Milky Way was itself a nebula among many. This suggested that some of the other nebulae he had observed were themselves collections of stars like the Milky Way. In this picture, the universe would not be a single collection of stars but rather was made up of many nebulae (what we would now call galaxies) that were 'island universes' – vast collections of stars.

Herschel was eventually proved right, but the Orion Nebula is not a galaxy – and Herschel also made it clear that more than one type of body had been given the name nebula. Just because things looked similar under the low magnification then available did not mean that they were the same type of phenomenon. Some were indeed 'telescopic Milky Ways' as Herschel referred to galaxies. But others were clouds of glowing dust that surrounded a star, nebulae that Herschel correctly labelled as stellar nurseries.

Herschel's idea that some nebulae were galaxies was ahead of its time, so much so that he himself eventually changed his mind and reverted to a picture of the Milky Way as the whole universe. This would be scientific orthodoxy right up to the early 20th century. It's notable, for instance that when Agnes Clerke, an Anglo-Irish astronomer, wrote her popular 1905 book *The System of Stars*, she commented: 'The question whether nebulae are external galaxies hardly any longer needs discussion ... No competent thinker, with the whole of the available evidence before him, can now, it is safe to say, maintain any single nebula to be a star system of co-ordinate rank with the Milky Way.'

We now know that Herschel's original wild speculation was correct, and there are vast numbers of galaxies outside of the Milky Way, but, equally, many nebulae really are hydrogen and dust clouds within the Milky Way that surround new stars in different stages of development. Stellar nurseries. And one such is the nebula in Orion.

Set course for the heart of Orion

Take a look at the constellation of Orion, a familiar pattern of stars from the night sky of the Earth.

View 6

The constellation of Orion

To see outside the ship, use the QR code or
visit www.interstellartours.co.uk/view6.html

What you are seeing on the viewing wall is a simulation – it includes a part of the night sky from Earth that will be familiar to many of you, along with the way the key stars relate to the traditional image. Like most constellations, the picture it supposedly represents takes considerable imagination to envisage. Orion is named for an Ancient Greek mythical hunter who was a distinctly dubious character. (Not only was Orion a rapist, he set out to destroy every animal on Earth. It's a strange choice for a hero.)

Next time there's a dark, clear night back on Earth, have a look out for Orion. Unlike many other constellations, it can be seen in both the northern and southern hemispheres, fully visible in winter in the north and in summer in the south. There is something about the pattern of Orion, particularly that distinctive combination of the belt of three stars and the outer four big stars, that makes it jump out to the eye.

It's worth including a reminder that the stars in a constellation are not related to each other in any way. The pattern is simply one that reflects what the stars happen to look like from the surface of the Earth. Out here in space, it soon becomes very obvious that they are not even the same distance from Earth. Of the bright stars in Orion, the most distant, Alnilam, the middle star of the belt, is 1,342 light years from Earth, while the nearest, Bellatrix, the top right of the four outer bright stars is just 245 light years away.

However, there's another bright object in the constellation that isn't a star at all, which forms part of Orion's sword – this is the Orion Nebula, even farther away from Earth at something like 1,500 light years distant*. This sounds a long way to go for our first significant stop, but it's the closest known stellar

---

* The distance from Earth to the Orion Nebula used to often be given as around 1,350 light years, but both NASA and the ESA put it at 1,500 light years.

nursery to Earth. Positioned visually roughly central in a triangle formed by the mid-belt Alnilam and the two bottom bright stars, Saiph and Rigel, sits the blob that is the Orion Nebula, also known as M42 from its designation in Messier's 1774 catalogue of 110 astronomical objects, or NGC 1976 from the *New General Catalogue of Nebulae and Clusters of Stars*, covering 7,840 items and assembled by John Dreyer in 1888.

It's always fun to speculate on pre-scientific mentions of astronomical features, and it has been suggested that the Great Nebula in Orion (to give the nebula its full original title), was first noted in historical times as part of a creation myth of the Central American Maya civilisation. This myth is only known from the period of contact of the Mayan people with the Spanish in the 16th and 17th centuries – it's not known how far back the story dates. It describes how all creation originated from a hearth, represented by a triangle of stars in the Orion constellation with the nebula at its centre. The fuzzy nebula was apparently supposed to represent the smoky fire. This seems an unlikely image, given the small size of the nebula in relation to the triangle – it feels like a modern reinterpretation, driven by the wishful thinking with which some like to highlight ancient wisdom.

We do know that the nebula was mentioned by the French astronomer Nicolas-Claude Fabri de Peiresc in 1610 and has regularly been described ever since. But given the naked-eye visibility of the nebula from Earth – even unassisted, it appears a little fuzzy if you have good eyesight – it seems strange that (Mayans aside) there aren't better-known early observations. This is the brightest nebula in the night sky. Some speculate that this is because the nebula has become brighter as it has been warmed up by the increasing number of stars inside it – if so, then the Mayan idea of it as the fire in the stellar hearth seems even more unlikely.

The stellar birthplace

We're just arriving at our viewing distance of the nebula, so we can switch to the real-time view. To make the image clearer, we're making infrared light visible here. The three bright stars in a row form part of a group of six in a relatively close cluster, four of them forming two binary systems. These are young, fully formed stars that have originated within the nebula, a grouping known as the Trapezium from the shape they make as seen from Earth. We can't quite see the whole nebula from our position, but it stretches around twenty light years across and contains somewhere between 1,000 and 2,800 stars in different stages of formation.

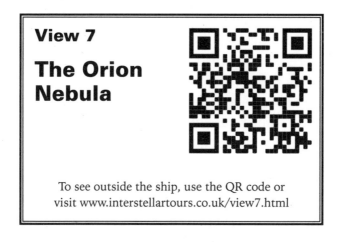

View 7

The Orion Nebula

To see outside the ship, use the QR code or visit www.interstellartours.co.uk/view7.html

We now know that the highly visible shape of the nebula is part of a much larger cloud of gas and dust known as the Orion molecular cloud complex, which is so big that when seen from Earth it extends beyond the limits of the whole Orion constellation. This is hundreds of light years across and includes a number of other separately identified nebulae, including the Horsehead Nebula, the Flame Nebula and M43.

The essential ingredients for making a star are these clouds of matter (primarily hydrogen gas), while gravity provides the cosmic equivalent of kitchen equipment. When atoms and molecules first formed around 380,000 years after the big bang, the majority of matter was in the form of hydrogen. Over the 13.8 billion years since, hydrogen gas was spread across the expanding universe, but during these vast periods of time, the tiny gravitational attraction between hydrogen molecules resulted in the formation of clouds. Where there was a slightly greater concentration, there was a positive feedback effect, resulting in more and more hydrogen molecules being pulled into relatively close proximity with each other.

One side effect of the gravitational squashing together of hydrogen molecules is that the gas in the cloud heats up. Compressing a gas always does this – think, for example, of repeatedly pushing the plunger of an air pump when inflating a bicycle tyre. Eventually, as gravity drags the components of the cloud in, there is so much hydrogen concentrated into such a small space (relative to the size of the overall nebula – stars are still enormous by our usual estimation) that there is enough heat and pressure for a nuclear fusion reaction to begin.

There are two ways to get energy out of atomic nuclei, the tiny, concentrated blobs of matter at the heart of atoms. One is when nuclei of large atoms split into parts. This is nuclear fission, the mechanism used in the nuclear power stations of the 20th century and the first half of the 21st. The other is nuclear fusion, when two or more small nuclei are joined together to make a new element – this process also releases energy, and it is the power source of the stars.

Fusion is extremely difficult to make happen (which is why working fusion power stations did not come online on Earth until the 2050s).

The need to produce intense pressure and heat before fusion occurs is why stars are so big. The Sun, for example, contains over 99 per cent of all the matter in the solar system. Stars need to be big because the more matter there is, the bigger the gravitational force squashing the nuclei together. The fusion process in most stars takes hydrogen, the lightest of the elements, and produces helium, which is the next lightest. This isn't a straightforward process as helium effectively requires four hydrogen nuclei to be combined – and an indirect route is taken to get that to happen – but there is so much matter in a star that the Sun, for example, fuses around 600 million tonnes of hydrogen each second.

Fusing for energy

The reason that it's so difficult to make fusion work is that atomic nuclei all have a positive electrical charge. A hydrogen nucleus consists of a single, positively charged proton particle. When two particles with the same electric charge are brought close together, they repel each other (just as your hairs repel each other and stand on end if you come into contact with static electricity). The closer the protons get, the stronger the repulsion forcing them apart. Yet to make fusion happen, the particles must be incredibly close together. This is because the force that holds together nuclear particles, known as the strong nuclear force, only operates over extremely small distances. This force is eventually far stronger than the electromagnetic force that is repelling the particles – but the separation between particles needs to be virtually non-existent.

In practice, even the massive temperature and pressure at the heart of a star is not enough to make much fusion happen. Without a remarkable extra ability of quantum particles, such as protons, stars would not work. The only reason we have the heat and light of the Sun to make life on Earth possible is a strange effect arising from quantum physics, the science of the very small particles of matter and light. Unlike objects that we are familiar with handling, quantum particles don't have a specific location until they interact with another particle. They exist as a cloud of probability – and this probability gives them the chance of being on the other side of an impenetrable barrier without ever passing through it.

This process is known as quantum tunnelling, a distinct misnomer as the particle doesn't make a tunnel through a barrier, but rather appears on the other side of it. Practically speaking, the repulsion between the protons of hydrogen nuclei provides just such a barrier. But the probabilistic nature of the location of quantum particles means that those 600 million tonnes of hydrogen nuclei each second in the Sun have materialised close enough to each other for the fusion process to occur.

With potential lifespans of billions of years, most stars maintain a balancing act. Two competing forces are constantly battling against each other in these vast nuclear furnaces. One is the nuclear reaction at a star's core, which generates energy in the form of light particles. This energy fluffs the contents of the star up, pushing matter away from the core. But this outward pressure is countered by the force gravity (without which those nuclear reactions wouldn't occur), pulling the star's contents inwards. For much of the time, this is an effective self-moderating system. If gravity starts to win, compressing matter more, then the rate of fusion increases, providing more outward pressure and reducing the compression. How long this balancing process can go on depends on how much matter went into the star in the first place.

Counterintuitively, the more mass a star starts off with, the shorter that its functional lifespan will be. Truly huge stars, which generate a vast amount of gravitational pressure, consume all their hydrogen in a few million years, while smaller stars can take many billions to work through it. Note, though, that all stars will eventually consume their hydrogen. Back in the 21st century, they used to refer to solar energy as 'renewable'. In reality, the energy reaching the Earth is a finite resource, but it does have a considerable lifespan ahead of it still.

The Sun's output is also more than we are ever likely to need on Earth. The energy produced by our neighbourhood star is immense, coming in at round 400 billion billion megawatts. To put that into context, just 89 billion megawatts of this hits the Earth – yet that was still more than 5,000 times our global energy consumption in the 2020s. Since then usage has plateaued as more efficient technology was introduced.

A familiar star

The Sun has been active for around 4.6 billion years already and is about halfway through its time as a hydrogen-consuming star, a state known in the astronomical trade as a 'main sequence star'. At the moment, its makeup is still around 73 per cent hydrogen. Despite this relatively long life, the Sun is a large star – in the top 5 per cent by size – which shows just how big the really fast-lived equivalents are.

Incidentally, ask anyone to draw a picture of the Sun, and they will almost certainly get its colour wrong. The light coming from the Sun is white – we just see it as yellow because some of the light spectrum from the Sun, particularly the blue end, is scattered by the Earth's atmosphere (this scattered blue light is why the sky appears blue). With the blue end removed, the effect is for the Sun to look yellow, or even red when it is low and its light passes through more air, scattering more of the photons.

Despite its whiteness, the Sun is a designated a yellow dwarf because yellow is the strongest of the colours in the light spectrum it produces, something that isn't obvious when looking at it from space, without air to scatter the light. Not that you should look at it directly in any circumstances. The Sun's light is so strong that it very quickly causes serious damage to the eyes. This is true even when the Sun is partially obscured – hospitals get a steady stream of people with permanently damaged eyes who have looked directly at the Sun during solar eclipses when the Moon is passing in front of the Sun.

Stars, then, are initially in the business of turning hydrogen into helium. (The very name 'helium' refers to the Sun's Ancient Greek name Helios, as the element was first discovered in the Sun, long before solar probes, using a telescope to detect the light signature of the element spectroscopically*.) This isn't how all the helium in the universe was made, as a little came from the post-big bang reactions of the early universe, along with a smidgen of the next heaviest chemical element lithium. But most of our helium and lithium, along with every other chemical element in existence, was produced directly or indirectly by stars.

---

* Spectroscopy makes use of the fact that all elements absorb particular colours of light when it passes through them or give off these particular colours when heated. This makes it possible to determine the makeup of stars without ever coming close to them (which is not a bad thing).

It might seem likely that with so many stars in so many galaxies, consuming hydrogen for billions of years, that there wouldn't be much hydrogen left for new stars to form in the future. But even today, 13.8 billion years into the life of our universe, around 73 per cent of the mass of the matter that makes it up is hydrogen, with about 25 per cent helium and the remaining 2 per cent accounting for everything else. That's come down from about 75 per cent hydrogen and 25 per cent helium when matter first formed after the big bang.

When a star has consumed most of its hydrogen*, it restructures to compress helium sufficiently for further fusion reactions to occur, producing some of the heavier elements, notably carbon and oxygen. Not every reaction will happen in every star, but all the elements up to iron have been produced this way. At this point, there's a problem in the stuff-making process. Where fusing nuclei up to this point release energy, to get iron to fuse to produce a heavier element requires energy to be put in from the outside. At this point, stellar fusion is no longer a self-sustaining process. If nuclear fusion in stars were the only way to produce heavier elements, we would only have 26 in the periodic table. No nickel, no copper, no iodine, no gold or silver or platinum (to name but a few). We'll have to wait for a later part of our voyage to see where these other elements came from.

From stars to solar systems

Although there are plenty of stars presently forming in the Orion Nebula, this is a process that will eventually run out of steam. If

---

* It's not uncommon to say that a star 'burns' hydrogen – but this is misleading, as burning is a chemical reaction that involves combining elements with oxygen, while nuclear fusion is a totally different process.

we look close to the bright stars of the Trapezium – young stars which are only around a million years of age – we can see that they are already reducing the likelihood of other stars forming nearby. Once active, stars blast out material in all directions in a process known as the stellar wind. This tends to clear the hydrogen and dust from around them, forming a kind of cavity within the material of the nebula. In these areas, no more stars can form.

However, some of the particularly young stars appear to have a disc of material around them – so called protoplanetary discs – which is where new planets are forming. The next step of our voyage will take us to a protoplanetary disc, a solar system in the making, and then on a tour of some of the galaxy's historically interesting planets, away from the more familiar names that are the planetary residents of our own solar system.

OTHER WORLDS

4

The nature of planets

As we head to a different part of the Orion Nebula to discover what a protoplanetary disc is, and to see one forming, we have a few minutes to think about just what it means to be a planet.

The familiar planets back in our solar system were originally thought of as 'wandering stars' – that's the origin of the word 'planet' in Ancient Greek. And for most of human existence, we were limited in our experience of planets to the most visible handful of bodies that orbit the Sun. The ancients knew of five: Mercury, Venus, Mars, Jupiter and Saturn. Strictly speaking, there were seven ancient planets, as the Sun and Moon also moved against the backdrop of the stars and so were classed as planets. But with the exception of the Earth itself, those five remain the only planets that are near enough to home to be visible with the naked eye.

The move to a model of the universe with the Sun, rather than the Earth, at the centre gave us a sixth true planet in the Earth itself. We then had the addition of Uranus, discovered in 1781 by William Herschel, followed by Neptune in 1846 and finally Pluto in 1930. This was before Pluto's ignominious down-grading to a minor planet in 2006, as a result of the acknowledgements that there were far too many similar bodies in the region beyond Neptune to consider Pluto to be one of the true planets*, leaving us with the current eight.

However, as soon as it was suggested that the stars were themselves distant suns, it seemed entirely reasonable that there could be other planets beyond those in the Solar System. When thinking about this, we need to discuss a man who is arguably the most misunderstood character in the whole history of science, the 16th-century Italian mystic, Giordano Bruno.

If you have read any popular science, or if you've watched any science history shows, you may have come across the character of Bruno as the number-one martyr of the scientific world. This characterisation is questionable, but whatever the truth, the man certainly was a character. Born in Nola in Italy in 1548 and burned at the stake in Rome in 1600, Bruno was a Dominican friar, a philosopher and a man bursting with ideas**. According to the frequently repeated story, his execution was the result of his suggesting that the stars were other suns, that those suns had their own planets and that those planets were teeming with life. There were, according to Bruno, many Earths – and this is said to have been considered

---

* Pluto's down-grading to minor or dwarf planet status was formally due to a failure to match the third of three conditions to be a Solar System planet: orbiting the Sun, massive enough to be rounded by its own gravity and clearing the neighbourhood of its orbit of other bodies. Technically, minor planets include all the asteroids, which don't match condition two, while dwarf planets only miss out on condition three. But arguably the best reason for Pluto's demotion is the one given above: that to include it as a planet would require us to add many more dwarf planets to the roster.

** Bruno has also become a great murder mystery detective in S.J. Parrish's excellent series of novels, starting with *Heresy*.

to be heresy by the Catholic Church, leading directly to his execution.

The truth is far more nuanced than this. Bruno was indeed executed for heresy – but for a very conventional heresy of the period: he argued against many Christian doctrines including the divinity of Christ, the virgin birth and the existence of the Trinity. This was more than enough to get him killed. Obviously, it was horrendous that Bruno should be executed for his beliefs, but this was common at the time. His apparently scientific ideas, while touching on concepts that were considered heretical, were not the cause of his demise.

It's also worth saying, as Bruno is often put forward as being the scientific genius behind early cosmological concepts, that his ideas were not original, nor were they even particularly scientific in nature. The notion of inhabited planets orbiting other stars dates back at least to the 15th-century German philosopher Nicholas of Cusa who wrote: 'Rather than think that so many stars and parts of the heavens are uninhabited and that this earth of ours alone is peopled – and that with beings perhaps of an inferior type – we will suppose that in every region there are inhabitants, differing in nature by rank.' Rather than be persecuted for his ideas, this same Nicholas was made a cardinal of the Roman Catholic Church.

As for Bruno's 'science', he rejected the observational and mathematical approach of the true early scientists of his period, instead relying on mystical pronouncements and waffly philosophical statements. So, with Bruno put in his place, we have the idea of inhabited planets around other stars going back at least to the 1400s. And as far as the existence of extra-solar planets goes, this is a concept that has since been proved right many times over.

A solar system in the making

We're arriving at one of the protoplanetary discs in the Orion Nebula – a new solar system in the making. Back at home, we're used to the system having a particular structure. There are smaller, rocky planets (Mercury, Venus, Earth and Mars) orbiting nearer to the Sun, and bigger, gas or ice planets further out (Jupiter, Saturn, Uranus and Neptune). To complete the picture, add in, for fine detail, a belt of asteroids between Mars and Jupiter and a collection of comets, dwarf planets, icy bits and more stretching far beyond the orbit of Neptune.

View 8

Protoplanetary disc

To see outside the ship, use the QR code or
visit www.interstellartours.co.uk/view8.html

It took a while to get a good explanation of why this particular cosmic arrangement had occurred. For example, it was speculated for many decades that the asteroid belt was the debris of a demolished planet that had been involved in a collision with another large body. In reality, the asteroids are bits and pieces of planetary building blocks that never managed to form anything like a planet. This was in part because of the disruptive influence of nearby Jupiter's gravitational pull, which constantly drags asteroids away from forming a stable structure, and in part because there simply isn't enough material in the asteroid belt to form a full-sized planet. So, let's get back to the protoplanetary disc in the Orion Nebula to see how it is possible to transition from a cloud of gas and dust to a fully functioning solar system.

The first thing to form is one or more stars. We are biased by experience with the Sun to think that only having one star in a system is the norm, but in practice, stars often come in pairs (binary systems) or even larger groupings. Some estimates put the number of stars in binary systems at around 85 per cent of the entire stellar population.

The star* sits at the centre of gravity of the disc of gas and dust that has formed around it and pulls more and more matter in. It's quite simple to see why we talk about a solar system rather than a planetary system – in terms of sheer mass, the star is the dominant component. In our solar system, for example, 99.8 per cent of all the matter is in the Sun. All the rest (including you, when back home) makes up just 0.2 per cent. Eventually, that central ball of gas becomes so massive that fusion occurs, transforming what had been simply a large collection of particles into a star.

Beyond the central star, that cloud of matter that forms the beginnings of a protoplanetary disc will not have been perfectly evenly spread out in three-dimensional space. There will have been areas with a greater concentration of gas and dust – these areas act as foci for the growth of planets. Initially, this would be due to direct contact, as grains of matter bang into each other and stick together, but over time, gravity will start to play its part as well. The more matter that is pulled in, the greater the

* Having said that binary systems are very common, for convenience of language, we'll just be referring to single stars in a solar system, but keep in mind that systems exist with more than one.

baby planet's gravitational pull on the material around it. Planets are a very natural accompaniment to stars and most normal stars have them.

Like pretty much everything in space, the original cloud is likely to have been spinning. This is simply due to the lack of symmetry in the material that makes it up. Unless everything started of beautifully evenly spread, as the material pulls in, it will start to rotate. The cloud would have originally been spread across three dimensions, but as the matter pulls in towards the centre, the natural rotation of that matter speeds up. As the material in the cloud spins around, it naturally tends to form into a disc shape, pulling the cloud into a plane oriented around the axis of spin. Think of a ball of pizza dough being spun on a chef's hand as the pizza is made. What starts as a concentrated near-spherical lump thins out to a shallow disc. In the case of the emerging solar system, this happens because there is a pull inwards towards the star, but there is not a gravitational pull at right angles to the direction to that central star, concentrating matter in the plane of the spin.

In the inner part of a solar system like ours, the Sun's heat and solar wind prevents the more prevalent, primarily gaseous materials from coalescing to form planets. Although heavier elements are much rarer than hydrogen and helium are, the heavier elements would have remained solid or liquid and bonded together to form the inner planets – though as a result of the material being rare, these planets would be relatively small. Get farther from the Sun and what we think of gases would be cold enough to form heavy gas-based planets like Jupiter and Saturn – effectively, these are failed stars in the sense that, like the Sun, they are primarily made up of hydrogen and helium, but are not big enough to start fusion. Even further out in our solar system, Uranus and Neptune still contain a lot of these gases, but also considerably more ice – typically solids composed of substances such as water, ammonia and methane.

Depending on the size of the star and how early the planets form, it is possible for planets quite close to the star to still have a sizeable hydrogen and helium atmosphere – as we'll see shortly. But first we need to move the ship to find a location further along the lifecycle of a star and its companions. The nebula has shown us the start of the process of forming planets, but let's see the finished product, visiting a few of the best-known exoplanets (the frequently used contraction for extra-solar planet). As we cruise to the first of our destinations, it's worth thinking a little about how these planets were first discovered. Although we can now go on an interstellar cruise like ours and go and take a look directly, remarkably, the first exoplanets were discovered at a time when humans had never travelled further than the Moon.

Tadmor, Draugr and Poltergeist

The obvious way to find a planet is to look for it. This is, after all, how we discovered most of the planets of our own solar system. (Neptune was originally found by its effect on the orbit of other planets, rather than by direct optical detection.) And spotting directly is now feasible from Earth for millions of exoplanets, but in the early days of such discoveries, direct detection was not possible. This is not surprising. Planets only shine by reflected light, and the amount of light energy received drops off with the square of the distance away from the source, meaning if, for example you double the distance, only a quarter as much light arrives. Compared to stars, planets are extremely dim and rapidly become more difficult to detect visibly as we look further from the Earth.

> **Speculation alert**
>
> As of 2023, only a handful of exoplanets had been directly detected, but the expectation is that with improved technology the number would grow quickly.

The first confirmed exoplanet was detected in 1992, and by 2023, more than 5,000 had been found. There is a whole range of possible mechanisms for detection, but the two most common are noting when a planet passes in front of a star, reducing its light output, and looking for a wobble in a star's position, caused by the variation in gravitational pull of a planet as it moves around its orbit, influencing the star's motion. In both cases, these methods work particularly well for finding massive planets: in the case of the wobble mechanism (more properly known as radial velocity), it's also helpful if the planet orbits close to its star.

We're not stopping off at the first *confirmed* exoplanets right now, as these were in a very unusual system, but we will come back to visit them later on in the tour. However, we are about to arrive at the first extra-solar planet that was detected*. Known as Gamma Cephei Ab, this planet was detected in 1988, before the first certain examples of exoplanets, but it wasn't confirmed until 2002. Although exoplanets are now usually given more friendly names such as Lich and Phobetor, they have a formal structured naming convention that looks a little odd, and

---

* A planet can be detected without being confirmed when it's either considered to be something else or there is not enough evidence to make it clear that it is a planet. Bear in mind that even Uranus was detected before it was confirmed, as Herschel first thought he had found a new comet.

certainly isn't as catchy as Vulcan, Trantor, Hoth or Qo'noS (the home world of the Klingons in *Star Trek*).

View 9

Gamma Cephei Ab

To see outside the ship, use the QR code or visit www.interstellartours.co.uk/view9.html

The first part of the planet's identifier is simply the name of the star it orbits. In this case, it's Gamma Cephei, located about 45 light years from Earth, and the third brightest star in the constellation Cepheus. This is a relatively obscure constellation of the northern sky, named after an Aethopian king in Greek mythology. Aetheopia was not the modern country of Ethopia but seems to refer to the area that is now primarily in Sudan. Its mythical king and queen were Cepheus and Cassiopeia, parents of Andromeda – all three accorded constellations.

Cepheus is roughly positioned in the upward direction from the rightmost stroke of the W of the better-known constellation Cassiopeia, about the spread of Cassiopeia's W away. The prefixes, alpha, beta, gamma and so on were given to stars within a constellation, intended to indicate the order of apparent brightness of the stars as seen from the Earth. In practice, these aren't always right as the measurement of brightness when these names were first applied was simply based on how bright they appeared to the naked eye.

Gamma Cephei also has a traditional name of Errai, from the Arabic for a shepherd. Before the discovery of its planets, Gamma Cephei's biggest claim to fame was that, as the Earth's axis gradually changes position against the night sky with time, it will replace Polaris as the pole star in about the year 3000. The letter 'A' in Gamma Cephei Ab applies not to the planet but is still a reference to the star. Gamma Cephei is a binary star, with Gamma Cephei A, an orange giant, the brighter of the pair. The other, Gamma Cephei B, is a red dwarf, which from the planet looks about twenty times as bright as the full Moon. Finally, the 'b' in the planet's name indicates that this was the first discovered planet of the star: subsequent ones would then be labelled 'c', 'd' etc.*

At least the name of the planet we are looking at is a little less clumsy than the first pair of confirmed exoplanets, known catchily as PSR B1257+12 A and PSR B1257+12 B, though they do now have semi-official names of Draugr (which were mythological Norse undead creatures) and Poltergeist. To makes matters even more confusing, the original scientific names were awarded before the development of the convention of naming exoplanets with the star name followed by a lower-case letter, starting at 'b' for the first discovered – so the first pair are *also* confusingly known as PSR B1257+12 b and PSR B1257+12 c.

Gamma Cephei Ab remains the formal astronomical name of the planet we're looking at, but it too has been awarded a more friendly name: Tadmor. This was the ancient name of the city of Palmyra in modern-day Syria, making it the astronomical

---

* The first discovered planet of a star is labelled 'b' because in the naming system the 'a' is considered to be the star, even though this letter isn't actually used. No, it doesn't make any sense, but that's not uncommon with astronomical naming conventions. Bear in mind that astronomers call all elements except hydrogen and helium 'metals'.

equivalent of the way that various European towns and city names also crop up in, for example, the USA and Australia.

As you can see, Tadmor is a very big planet – it's nearly ten times as massive as Jupiter, making it even closer to being a star. Tadmor would only have needed a little more mass to have been able to put out some energy as an extremely faint brown dwarf star. It orbits close for such a massive body – if it had been in our solar system, it would be just outside the orbit of Mars. Tadmor was detected from Earth using radial velocity – measuring the wobble of the star. This is not discovered by seeing the star shift in the sky – the movement would be too small to pick up; however, when the star is moving towards us in its wobble, its light becomes a little more blue (it is blue-shifted) and when it moves away, the light becomes a little redder (red-shifted) – a change that proved measurable.

As you can also see, as a gas giant, Tadmor is not the kind of planet where we would expect any form of life to have developed, nor is Gamma Cephei A much like the Sun – and while we certainly don't know that it's necessary to have a Sun-like star to support life, this is the only type of star where we know for certain that life can develop, thanks to our own existence.

Let's now take in a planet that initially seemed more likely to be a home for life. After jumping back to Gamma Cephei from the distant Orion Nebula, it's a good thing that our drive system takes exactly the same time to take any jump within its range, as we're heading a long way out to Kepler-62, a star that is around 990 light years from Earth and located in (but not considered part of) the constellation Lyra as seen from home.

The Goldilocks zone

The name Kepler-62 comes from the Kepler probe, a specialist space telescope that was in operation between 2009 and 2018.

Kepler, named after the 16th-century German astronomer Johannes Kepler, who devised the laws of planetary motion, was specifically designed to detect exoplanets by the transit method, spotting stars that dimmed in a regular way as planets passed in front of them, reducing the light that came towards the Earth. Kepler-62 has five planets, two of them sitting in what is sometimes called the Goldilocks zone – the distance from the star where it's neither too hot nor too cold for liquid water to form.

It's hard (though not impossible) to envisage life existing without water or a similar fluid (such as liquid hydrocarbons at far lower temperatures) to enable biological processes to take place. This is because all life that we know of relies on mechanisms operating at the scale of large molecules. Think, for example, of DNA and the molecular machinery that produces proteins based on the genetic code. At this scale, electromagnetic forces make it easy for dry-matter particles to stick together – it is only with a substance such as water acting as a lubricant that the molecular machinery can function.

The star Kepler-62 itself is both cooler and smaller than the Sun – it's an orange star, but it is not too large for its size to be a problem as far as planets are concerned.

View 10

Kepler 62f

To see outside the ship, use the QR code or visit www.interstellartours.co.uk/view10.html

As it happens, the most interesting planet to be discovered was the outermost one, Kepler-62f, because it was both in the habitable zone and likely to be rocky (as we can now see it is, since we have moved into orbit around it). Since Kepler-62 gives off less energy than the Sun, its habitable zone is closer than the one the Earth sits in – Kepler-62f is roughly the same distance from its star as Venus is from the Sun. It is what's known as a super-Earth: the planet is bigger than Earth but smaller than the ice giants in our solar system. It's about three times the mass of Earth – so any inhabitants would face some serious gravity – heavy going for us, but not impossible for life to exist. It's about 1.4 times the diameter of Earth in size. When it was discovered, this size and mass suggested that it was rock, and it was initially thought that it may have had oceans.

However, we now know that the planet has no atmosphere.

Speculation alert

It was known in the 2020s that Kepler 62f does not have an atmosphere and has no magnetic field – but equally, we didn't know it did.

Without an atmosphere, there's a real problem for habitability, leaving aside that most of the life we are familiar with requires an atmosphere or oceans carrying oxygen. If the Earth had no greenhouse effect from greenhouses gases in the atmosphere, the average temperature would be $-18°C$ ($0°F$). That would mean no liquid water and little chance of life. Kepler-62f is on the edge of its star's habitable zone. Without any greenhouse effect, the surface temperature is much colder, around $-65°C$ ($-85°F$), not unlike Mars before it was terraformed.

> **Speculation alert**
>
> Terraforming is the act of making a planet more Earth-like, so that it is suitable for human habitation. It has often been speculated that Mars may at some time in the future be terraformed to provide more living space in the solar system, although some believe it will never be viable.

It's possible the planet had an atmosphere once, but unlike the Earth, it doesn't have a magnetic field to keep away the stream of charged particles from its star that forms the stellar wind: as a result, the atmosphere would have been stripped away over time.

Meet the neighbour

Our penultimate planetary destination (for now) is the first exoplanet ever visited.

> **Speculation alert**
>
> As of 2023, it wasn't certain which would be the first extrasolar planet to be visited, should humans ever manage interstellar flight. But the planet mentioned here is an obvious early target because of its closeness to our solar system. Similarly, details about the planet would have been speculative, although they are based on reasonable assumptions from what was known in 2023.

One of the benefits of our drive system is that we don't have to see things in sequence in the order they occur as we move away from Earth – and now we're back in our stellar backyard, just a few light years from home, near the star Proxima Centauri, taking a look at Proxima Centauri b (or if you want to go for the full formal naming convention, Alpha Centauri Cb). This was never expected to be a planet that was likely to harbour life – and it doesn't. However, its nearness to Earth did make it an obvious early stopping off point.

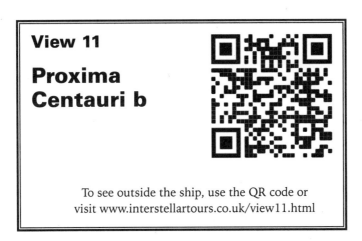

View 11

Proxima Centauri b

To see outside the ship, use the QR code or visit www.interstellartours.co.uk/view11.html

At first sight, this planet has a lot going for it. It's a rocky planet, similar in mass to the Earth, and it does sit within Proxima Centauri's Goldilocks zone. However, if you want life to survive, this is not a great star to orbit, nor does the planet have an ideal orbital situation. Firstly, it is tidally locked. This means that it is in the same state as Mercury in our solar system. One side of the planet always faces the star – this would mean extremes of temperature between the two sides of the planet. Secondly, the planet is also much closer to its red dwarf star than any planet in our solar system, at about 1/8 of the distance Mercury is from the Sun. This means that Proxima Centauri b is much more at risk from the star's volatility – and this is a very capricious star. In just a few hours it can vary in brightness by a factor of 100, sometimes blasting out large quantities of material in flares – its stellar wind is hellish and long ago would have stripped any atmosphere from the planet and destroyed any potential life-forms with its radiation.

The first signs

Leaving Proxima Centauri behind, we finish our planetary tour at a location with a near certainty for life, the planet known as Trappist-1 e.

View 12

Trappist-1 e

To see outside the ship, use the QR code or visit www.interstellartours.co.uk/view12.html

Speculation alert

This planet is one of several thought in 2023 to be among the most likely to be habitable of those then discovered, though many of the details were speculative at the time.

This planet itself is not dissimilar to Proxima Centauri b. Its mass is a little less than that of the Earth, but at around 80 per cent of Earth mass, it has sufficient gravitational pull to be comfortable to humans. It is a rocky planet, orbiting close to its dim, ultra-cool red dwarf star, situated about 40 light years from Earth and located in the Earth's sky in the constellation of Aquarius. The planet's year is just six days long, and it orbits just 4.5 million kilometres (2.8 million miles) from Trappist-1 – compare that with Mercury in our solar system, which is around

70 million kilometres (44 million miles) from the Sun. And, like Proxima Centauri b, it is tidally locked.

However, its host star is significantly more stable, and as a result it is a planet with liquid water oceans and an Earth-like atmosphere, far less troubled by its star than Proxima Centauri b. We have been investigating the planet very carefully for some time now. We know that there are no higher forms of life present on it, but it is possible from the atmospheric makeup that there is something alive down there. As yet, careful protocols have been employed to ensure that there is no contamination from Earth-based organisms, and because of this we are yet to have a certain detection of a living organism anywhere that isn't from our home system.

So where are all those aliens?

If we were to take science fiction literally, we might expect that these extrasolar planets would be teeming with life, but in reality, living organisms seem to be relatively rare in the galaxy. We certainly do not expect to find the typical 'human with a twist' aliens portrayed in many science fiction stories. It's important to remember that the beings in science fiction are usually created to help us understand how humans react to and are influenced by scientific discoveries and changes, including space exploration.

The role of aliens in, for example, *Star Trek*, was often to explore different aspects of human nature, rather than genuine alien concepts. The assorted Klingons, Vulcans and Romulans that were introduced in the series were intended to embody aspects of human traits, rather than truly alien species. Admittedly there were more distant aliens in the form of, say, small furry tribbles, or intelligent non-embodied energy – but even they reflected Earth-based animals or human thought processes without the biological frame.

Even in written science fiction, which tends to be more sophisticated than the TV or movie equivalents, it's relatively rare to find aliens that are genuinely... alien. And when it comes to discovering the real thing, the chances are that the vast majority of living things – if they exist at all elsewhere in the galaxy – are more likely to be the equivalent of bacteria rather than intelligent higher forms like human beings.

Even on the Earth, bacteria far outnumber the larger-scale organisms. On average, every human is host to around to 30 to 40 trillion other organisms in the form of bacteria, viruses and microbial fungi. And that's without looking at all the microorganisms in the Earth's soil, air, sea and in other animals and plants. Equally, for around 1.5 billion years, the only life on Earth was in the form of bacteria and other similar microbes. It would hardly be surprising, then, if most planets that were discovered with life only hosted something like bacteria.

In the past, there have been attempts to estimate the chances of there being higher levels of alien life that would be capable of interstellar communication. This was typified by the Drake equation, produced in 1961 by astrophysicist Frank Drake, who we've already met as a contributor to the Pioneer plaques and the Voyager discs. Unfortunately, the Drake equation proved to be rather like those phoney formulae that appear in the news every now and then. The kind of thing where someone declares they have the mathematical recipe for an ideal life or how to make the perfect sandwich.

The Drake equation multiplies together a whole string of estimated probabilities. Some have the potential for a degree of accuracy – such as the fraction of stars with planetary systems, something that was reasonably well established during the first part of the 21st century. But plenty of the probabilities were pure guesswork. There was (and still is) no scientific way to determine, for example, a factor such as the 'fraction of life-bearing planets on which intelligent life emerges' – because we only have the evidence of this happening on a single planet: Earth. It is not possible to extrapolate anything meaningful from a single data point. In the end, the Drake equation is a fancy way of saying, 'The answer could be anything.'

As yet, there is no good scientific evidence for technology-producing aliens existing. This could, of course, be simply down to the size of the galaxy and the limitations imposed by the light speed barrier. Without the development of our hyperspace drive, interstellar travel is likely to be local and unmanned. It is entirely possible, though, much though romantics would love the galaxy to be teeming with interesting lifeforms, that Earth is extremely unusual in having higher lifeforms – or even in having life altogether.

It's worth noting that life on Earth only appears to have emerged once. Every known living organism appears to be related. Similarly, all known higher life on Earth involving complex cells has a common ancestor. It was a one-off development, which implies it may be a rare occurrence. The hyperspace drive is new enough that we are yet to make sufficient inroads to be sure about the prevalence of life in our galaxy. But time will tell.

As we have seen, planets, unlike stars, require heavier atoms than those produced after the big bang to be part of their makeup. Even the gas giants like Jupiter in our solar system that still contain a lot of hydrogen also contain a range of heavier elements. These elements must have come from somewhere. And our next stopping point will bring us to the moment that many such atoms came into being. We are about to visit an astronomical calamity of vast destructiveness that is simultaneously the forge for essential new substances.

SUPERNOVA 5

As the walls of the starship fade away once more to show the depths of space, you might think that we have arrived back home. Ahead, once more, is a familiar pattern from the night sky of Earth that has been a central feature of our voyage so far – the distinctive constellation of Orion.

View 6

The constellation of Orion

To see outside the ship, use the QR code or visit www.interstellartours.co.uk/view6.html

Remember that what you are seeing in front of you is a simulation – it's a view of the night sky from Earth (along with the traditional mythical image). The other important thing to continue to bear in mind is that the constellations visible in the night sky are collections of stars and other astronomical bodies that are not linked in any way – they are not close to each other, and it is just a visual coincidence that they look this way from the Earth's surface.

Encountering beetle juice

In our earlier exploration of the Orion constellation, we were located by the nebula that is in the bottom half of the constellation. Now we are between a half and a third of the distance from the Earth that we were then, between 500 and 600 light years away.

Look up to the top left of the constellation (around Orion's left armpit in the mythical image) and you will see a bright red star, Betelgeuse – a name that is close enough to 'beetle juice' to have inspired the movie of that name. To us astronomers, its proper name is Alpha Orionis, but to be honest, we all still call it Betelgeuse. Delightfully, this word is the result of a mistake: it originated due to a misreading of its actual Arabic name Yad al-Jauza. The star was known, of course, long before it got that or any other name – it's hard to miss it in the night sky. Orion is one of the most widely visible constellations around the world, and Betelgeuse stands out as it's one of the few bright stars in the Earth's sky that has such a clearly distinctive colouration.

But let's switch to the actual view from outside the ship. This is Betelgeuse in all its fearsome glory, a vast, pulsing and roiling mass of a red star. We've added a graphic overlay to show the comparative scale of part of our home solar system.

View 13

Betelgeuse

To see outside the ship, use the QR code or
visit www.interstellartours.co.uk/view13.html

This, then, is what Betelgeuse looks like right now, outside the
walls of the ship. But it won't stay this way for long. This star
is about to undergo an explosive transformation as it becomes
a supernova. Forecasting the exact time at which this will happen is tricky – as the great 20th-century Danish physicist Niels
Bohr (and many others) once said, prediction is difficult, especially when it comes to the future. However, we can say with
98 per cent certainty that Betelgeuse will explode within the
next twenty-four hours. You are very lucky: no other tour, past or
future, will see this particular event live as you are about to do.

Speculation alert

Exactly when Betelgeuse will go supernova is unknown, but it is
likely to be within 100,000 years or so. In the early 2020s, the star
did dim significantly and then recover, which some suggested
was a sign of an imminent explosion, but astronomers think it
is more likely to have been due to a dust cloud or dark patches
equivalent to sunspots.

We expect that Betelgeuse will end its days in what's known
as a type IIP supernova. (That's 2P, with the '2' represented by

Roman numerals.) Very shortly, the star's core will collapse to form a neutron star, a fascinating type of body that we will have the opportunity to visit in more detail later, while the rest of the star will be blasted outwards to seed space with matter for light years around. Don't worry, by the way, about your safety. By partially engaging our hyperspace drive in a limited local oscillation, we can ensure that, without moving, there will be no possible damage to the ship.

I'll talk you through what is happening in more detail when the collapse begins, but while we're waiting, you ought to hear a little more about the whole concept of the supernova.

A new star

In ancient times, anything unexpected in the night sky would inevitably be referred to as a new star – in Latin, the European academic language of choice for centuries, that made it a 'stella nova'. This was something of a problem for the theory of astronomy that held for nearly 2,000 years based on Ancient Greek science. The assumption was that, beyond the rotation of the various heavenly spheres carrying the stars and planets, the heavens above the orbit of the Moon were unchanging.

Even the motion of the planets caused something of an issue. The original assumption had been that each planet (including the Sun and the Moon) travelled around the Earth attached to its own vast crystal sphere. But some planets, most dramatically Mars, appear to jiggle about in their path, even reversing their direction of movement. We now know that this is due to the relative motion of the Earth and the outer planets in their orbits, but it forced classical astronomers to devise more complex models involving spheres within spheres – so-called epicycles – to explain the complex progression that was observed.

This still left the problem of comets. These apparent new stars, sometimes equipped with portentous glowing tails, crossed the heavens over a lengthy period and then disappeared. The only way they could sensibly fit with the Ancient Greek model of science was if they were passing overhead lower than the orbit of the Moon, although it took a fair amount of imagination to fit comets into the big picture this way. Meteors – bodies from space dust up to large rocks that become trapped in Earth's gravity well and fall to become meteorites – might also have been labelled as new stars, although their period of visibility in the night sky is so short there was rarely any potential for confusion.

The hot news of 1006

Comets were the most common forms of stellae novae, but occasionally, a different kind of new star appeared. Some of these bright points of light in the sky were intensely bright – a few of them could even be seen clearly in the daytime – and although such a star would fade over time, it remained at a fixed point in the heavens, rather than carving out its own path like a comet. This seemed to place these particular stellae novae in the crystal sphere of the 'fixed' stars. We now know that they will have been supernovae. It's hard to be certain exactly which of the early accounts of new stars was the first true observation of a supernova to be recorded, but we can be certain that one spotted around the end of April in 1006, now thought to be the brightest such event in recorded history, was the real thing.

This was because the star's appearance was reported across the world – in China, Japan, Korea, the Middle East and Europe.

After seeing it, the Egyptian physician Ali ibn Ridwan told us that the new star was 'a large circular body, 2.5 to three times the size of Venus. The sky was shining because of its light. The intensity of its light was a little more than a quarter of that of moonlight.' Similarly, a chronicle from the Benedictine monks of St Gall Abbey in the north-east corner of Switzerland noted that the new star was 'glittering in appearance and dazzling the eyes, causing alarm. In a wonderful manner it was sometimes contracted, sometimes spread out, and moreover sometimes extinguished.' This behaviour suggests to us that this was a different type of supernova to the one we are about to experience – the 1006 incident appears to have been a type Ia. We will discover more about one of these in a moment to compare it with what's happening to Betelgeuse.

At an unknown point in history, the stella nova became simply the nova. Already by Newton's day, in the late 16th and early 17th century, Latin's use as the language of science was fading. It's likely the 'stella' part was lost in the 18th or 19th century. By modern times, novae were divided into three broad categories dependent on how bright they were: lower-class, middle-class and upper-class. And it was in 1932 that Swedish astronomer Knut Lundmark referred to the brightest, upper-class novae as super-novae, a name that stuck, although the hyphen would fade away faster than these new stars*.

That reminds me, I ought to clarify that 'ae' bit on the end of novae and supernovae, which you can pronounce 'eye' or 'ee' to taste (I prefer 'eye'), as no one has a clue about how Latin

---

* There is some controversy over who first coined the term 'supernova'. Everyone agrees that Lundmark was the first to publish a paper using 'supernova', but some claim that Fritz Zwicky and Walter Baade had already been using the term for a couple of years by then. As the only academic source for this claim is a paper by Zwicky, it is at the very best a questionable assertion.

was really pronounced. The term comes from the Latin plural of nova, although some argue that astronomers were being overly fussy about their classical roots and should have stuck to simply calling the phenomena novas and supernovas. While that argument has a lot of sense, it seems too late to correct this particular nominative aberration.

Searching for causes

In 1934, the Swiss–American astrophysicist Fritz Zwicky, then working at Caltech in the United States, joined the supernova scene at the same time as American astronomer Walter Baade at the Mount Wilson Observatory. Between them, the pair expanded the work on supernovae from astronomy alone to include astrophysics. Broadly, astronomers collect facts, while astrophysicists try to explain those facts using physical theory. All astronomers cared about back in the early days of describing them was how bright the various novae were. But Zwicky and Baade realised that supernovae represented a totally different type of astrophysical event to the fainter novae. What distinguished them was *not* their brightness, but the different possible physical events that might have caused the flare up, making them appear to be a new star as seen from the Earth.

Both novae and supernovae involve kinds of stellar catastrophe, but the deciding factor between the two, once the distinction was clearly made, was that a nova involved a relatively small part of the star exploding, leaving a smaller but similar star, while a supernova was a sufficiently catastrophic event to totally transform the star and could produce much more interesting remnants in its detonation.

With a number of exotic exceptions that we can ignore here, supernovae divide into three broad categories, although astronomers, who love putting things in theoretical boxes, then divide each of these into smaller subcategories. As we have heard and soon will see, Betelgeuse is about to become a type II supernova. These are stars that have plenty of hydrogen remaining, and it is the collapse of the star's core that causes a vast explosion. The distinction that specifically makes Betelgeuse a type IIP simply reflects the way that the luminosity of the supernova is expected to tail off with time – 'P' stands for 'plateau'. Supernovae of this kind tend to go through an extended period of about 100 days after their explosion with a relatively consistent, bright output of light.

By contrast, the Ia supernovae (and a number of related variants), which tend to be particularly bright, have a totally different mechanism. These are the type most frequently seen from Earth simply because of their brightness, even though there are roughly half as many of them as there are there are type II supernovae over a period of time. That excessive brightness is mostly due to the decay of an isotope of the element nickel, produced in the initial explosion. Type Ia supernovae occur in stars that have pretty much entirely consumed their hydrogen and helium content via a rather complex route, which we'll see in a moment. Finally, there are the other type I supernovae, mostly Ib and Ic variants, which have some hydrogen left but still undergo a core collapse similar to the type II – the processes are subtly different, but the basic supernova mechanism is the same and the results are similar.

The end of the Sun

Let's start with thinking about a star that won't quite make it to supernova at the end of the initial period of its life, but please be aware we may have to switch back to Betelgeuse at any moment if things are on the verge of kicking off. This star is one that has become a red giant – the fate of our own Sun after it is has run out of hydrogen in the core and started to fuse helium. There is still plenty of hydrogen in the outer layers of the star, but at the core, all the hydrogen has been consumed and it is helium that is now fusing. As a result of the changes at the core, the outer layers of the star have fluffed up, making those outer parts both bigger and cooler than they previously were – hence it being a red giant.

We obviously can't show you the actual Sun after this happens, but the diagram on the viewing wall provides an idea of the scale that would be involved.

View 14

Sun after core hydrogen exhaustion

To see outside the ship, use the QR code or visit www.interstellartours.co.uk/view14.html

One crucial factor in what will happen next is size. The Sun's fate is what we might expect for similar stars up to around eight times the Sun's size. (We'll get back to the bigger ones later.) As the red giant comes towards the end of its life, the helium at the core is pretty much consumed, but even as the outward pressure from this drops off, there isn't enough gravitational pull (hence the importance of the size factor) to start the next level of fusion, using the carbon product of the helium fusion. The core is being restrained from collapsing any more by a quantum effect known as degeneracy pressure, produced by the electrons that have been stripped from the atoms, fighting back against further collapse.

Electron degeneracy is a quantum mechanical effect. There are two broad types of quantum particles. Some, like photons of light, are called bosons. They can be packed together into the same space without any problem. Photons simply ignore each other, passing through each other as if they were not there. But the other type of particle, known as fermions (such as electrons), obey something called the Pauli exclusion principle. This sounds like it should be a descriptive concept, but in reality, it is the result of a force of nature. Fermions can't be part of the same quantum system (an atom, for instance) if they have a set of identical properties. There always has to be something different about them or they aren't allowed in.

When packed close together – as is the case when the core of a star collapses – fermions push back. In a star's core, the electrons won't be part of atoms, which might make it seem that the exclusion principle doesn't apply, because they aren't part of the same atomic system. However, there's a weirdness about quantum particles that means that there can still be a problem*.

If I have what you might think of as a visible version of a particle – a tennis ball, say – I can specify exactly where it is. If I throw it, under normal conditions, simple equations will enable me to calculate its exact trajectory. Quantum particles aren't like that. Unless they're interacting with something else, they don't

---

* There was a trend in the 2020s among some science writers to suggest that there is nothing weird about quantum physics. In a sense this was true – quantum theory is a description of how reality works, so really it's the natural thing. But because it works totally differently from the way visible objects work in our day-to-day experience, it seems entirely reasonable to still describe it as weird.

have a specific location, but rather exist as a collection of probabilities of being in different locations. We can't say anything else about where they are or how they get from A to B. This means that if we have a whole bunch of electrons, squeezed very close together, at any one time, many of them have a probability of being in any particular location – and that means they bump up against the Pauli exclusion principle.

The more pressure that is applied and the closer packed the particles, the more they resist being together. One of the properties of a quantum particle is its energy level – this is quantised – it can only have one of a specific series of values rather than a continuously variable value. Coins, for example, are quantised. If your smallest unit of currency is, say, 1 cent or 1 penny, despite what accountants might tell you, you can't have a 0.23 cent or penny piece. Similarly, a quantum particle's energy levels in a system can only have certain incremental values. As electrons are crammed together in the stellar core, the lower energy levels of electrons are already taken up and electrons are forced to exist at higher and higher levels. This increased collective energy is reflected in a pressure that resists the inward push of gravity.

As a star, like the future red giant Sun, reaches a certain point, we're left with a white-hot remnant at the core that is blasting out high-energy photons. There is no more fusion occurring, but the star keeps glowing due to its residual heat. Although eventually such white dwarfs* are expected to fade to black, no such 'black dwarfs' are yet to be detected, as it's expected that it will take more than the universe's current 13.8-billion-year lifetime for the glow to fade entirely.

---

* I know the plural of dwarf is dwarves – please don't contact to tell me I've got it wrong – but as we have seen, astronomers have what might politely be called a distinctive way with words. And they use 'dwarfs' as the plural for dwarf stars whether we like it or not.

As we saw with the Breakthrough Starshot project, light exerts a pressure – this is what fluffs up a star in the main sequence of its life – and when a red giant ends up with a white dwarf core, these high-energy photons have enough momentum to push the outer parts of the star away from the remnant star. We end up with two distinct parts. The white-hot core, often no bigger than the Earth, is the white dwarf, while the surrounding patch of glowing material, kept energised enough to give off light by those high-energy photons, is known as a planetary nebula.

Planetary nebulae come in a whole range of shapes and sizes – we can't know for sure what the Sun's planetary nebula will be like. But we're showing the Helix Nebula on the viewing wall, pictured here using a whole range of different light detections. It is located in Aquarius, about 650 light years from Earth.

View 15

Planetary nebula

To see outside the ship, use the QR code or
visit www.interstellartours.co.uk/view15.html

This outcome is what we expect for the Sun in a few billion years' time. But we haven't got ourselves a supernova in this case. There was no explosion. We just see a gradual fading away into old age as the nebula dims and disperses, and over many billions of years the white dwarf cools and goes dark.

Why would it go bang?

It seems there needs to be something more happening to get a supernova. We know how a type Ia supernova occurs, but as yet we are not always sure of what triggers it. In the case of such a supernova, something – let's not worry what yet – has caused the white dwarf to experience so much inward pressure that it can overcome the outward degeneracy pressure and carbon begins fusing. Without the outer parts of the star to act as a balance – this is a pure stellar core – the fusion process runs away, faster and faster. In effect, it becomes the carbon equivalent of a hydrogen bomb, a thermonuclear explosion, but a nuclear explosion involving an amount of compressed matter the size of the Earth.

How can this come about? One way, as we will see in our simulation, is to introduce a second player – another white dwarf. This is not as unlikely as it seems. As we have seen, although there are plenty of standalone stars like the Sun out there, there are many more multi-star systems, most commonly binaries made up of two stars. This isn't surprising when you think about the way that stars form. Although it's entirely possible that most of the gas and dust will accumulate to a single object, there is no reason at all why there shouldn't be two competing centres of attraction, and it's likely that two or more stars will form.

We now have what appears to have been the starting point for the ultra-bright supernova that was seen on Earth in 1006. Many astronomers now think that this was most likely to have been the product of a pair of white dwarfs colliding. As we watch the white dwarfs dance around each other, their mutual orbit is decaying, disturbed by the formation of these stars from cores of their predecessor red giants. Before long, they will collide in a stellar spectacular. This provides enough extra energy and gravitational pressure to start that runaway carbon fusion ... and we get the expected explosion.

It might seem that this could be prevented from happening by the resistance provided by the electron degeneracy pressure. But there is a natural limit to this. As the pressure increases, the speed with which electrons move around goes up. There comes a point where that speed approaches the speed of light. Electrons can't move any faster than this – if a collision results in an even more extreme compression, even the Pauli exclusion principle gives way. There is nothing left behind that resembles a star. The whole body has been blasted out into space, producing the dramatically bright burst of light that is typical of a type Ia supernova. (For your comfort and safety, the light levels will be drastically reduced in the simulation.)

View 16

Type Ia supernova

To see outside the ship, use the QR code or visit www.interstellartours.co.uk/view16.html

A similar situation could also happen with a single white dwarf and a red giant that hasn't reached the same stage of development. Either possibility could occur in a binary star system – both stars could be very similar, or one could be different in size and take longer to form and begin fusion. With such an unbalanced binary, when the larger red giant puffs up, its outer matter could come close enough to the white dwarf for the smaller star to begin pulling matter off its sibling, increasing its mass until carbon fusion begins, but not providing enough of a structure to prevent a runaway process.

When visiting Orion for the first time, we discovered how some stars produce elements up to iron in mass, but in the type Ia explosion, we are starting with a star that has only produced carbon and oxygen as its highest-level products. There is such energy in the explosion, though, that studies of the remnants of these stellar explosions show both the elements above oxygen up to silicon and another group around iron. This still doesn't give us the many heavier elements we find in the universe, but it gives a first hint of how other supernovae may provide the rest.

Back to Betelgeuse

We've just heard that Betelgeuse is showing signs of enhanced activity, which make this the ideal time to switch back to a real-time view of what's outside the ship. Betelgeuse is a red supergiant with a mass of at least fifteen times that of the Sun. Unlike the relatively gentle evolution of a smaller red giant, by the point of its life we see it at now, the core of Betelgeuse will have gone through several stages of fusion, producing heavier and heavier elements. Now it has gone all the way through the fusion processes, producing carbon, then neon, then oxygen, then silicon and finally iron.

As we have seen, fusing iron cannot occur without putting energy into the system – the process can go no further. But the immense pressure produced by gravity in a huge star like Betelgeuse means that when fusion stops, the core collapses until what is left is a relatively tiny ball of neutrons – a remnant that is perhaps only twenty kilometres (12.4 miles) across despite containing the mass of one to two Suns. Compare that with the size of the Sun, at around 1.4 million kilometres (870,000 miles) in diameter.

Neutrons are electrically neutral particles (hence the name) found in the nuclei of atoms. Unlike protons, their positively charged companions in atomic nuclei, neutrons can pack together to form an incredibly dense substance. Ridiculously so. Just one teaspoonful of this neutron star material has a mass of around 100 million tonnes. That's about 300 Empire State Buildings, twenty pyramids of Giza ... or around 20 trillion teaspoons of sugar.

There's an immense gravitational pull from such tightly compressed matter. If you dropped an object onto a typical neutron star from a height of one metre (3.3 feet) it would hit the surface at around 1,930 kilometres (1,200 miles) per second. That's why we can use neutron star material to produce gravity on the ship. But inside the dying star, the collapse stops abruptly when the neutron's equivalent of the electron degeneracy pressure, mentioned in looking at the Sun's future, cuts in.

As the collapse suddenly and catastrophically ends, the core bounces back, sending out an immense shockwave that blasts away the outer parts of the star.

View 17

Type II supernova

To see outside the ship, use the QR code or visit www.interstellartours.co.uk/view17.html

This is a far bigger and faster expansion than the gentle drift outwards of the planetary nebula at the end of a Sun-sized red giant. There is some dispute over exactly how a neutron star forming supernova works – originally it was thought that the core collapse and bounce was enough to do the job, but later modelling suggested such a shockwave would fizzle out before it could blow off the outer layers. It's possible, though, that the huge production of particles called neutrinos in the formation of the neutron star could give the extra momentum needed.

Neutrinos are very low-mass particles produced by nuclear reactions. They only interact faintly with other matter – most neutrinos that come streaming from the Sun, for example, pass straight through the Earth. However, the production from the formation of a neutron star would be so vast that there may be enough neutrinos to give the shockwave sufficient boost to complete the supernova effect.

The energy produced in the collapse is sufficient to produce more fusion, creating elements that are heavier than iron, some of which are radioactive and decay. Along with the light emitted by the intensely hot ejected material, the photons emitted from this nuclear decay give the supernova its intense luminosity. In the material streaming away from Betelgeuse, the sudden, dramatic compression produced by the shockwave has fashioned many of the other elements heavier than iron that will eventually end up in planets such as the Earth. This is the conception moment of what will eventually be rocky planets and everything they will contain.

The atomic foundry

Let's just underline what's happening as Betelgeuse meets its end. The big bang left us with hydrogen, helium and a bit of lithium. Pretty much every bit of hydrogen in your body, for example, dates back to a few hundred thousand years after the big bang, around 13.8 billion years ago. But if that were all there had ever been, you would not exist, nor would our starship. There would be no solid stuff to speak of. Your body is primarily made up of eleven elements: apart from the hydrogen, the other ten were made in stars and spread by supernovae. We (and the Earth) are products of supernovae.

Some have also suggested that it would be possible for a supernova at some point in the future to kill off a fair amount of life on Earth. A relatively near supernova would certainly be a danger to life. There would be a direct impact from radiation, plus that radiation would likely strip away the ozone layer, allowing far more of the Sun's destructive ultraviolet through, causing cancers and mutations. Some elementary detective work suggests that one or more supernovae exploded relatively near Earth about 2 million years ago.

We know this because a signature product of supernovae is the iron isotope iron-60. An isotope is a variant of an element with different numbers of neutrons in the atomic nucleus. The most common and stable isotope of iron is iron-56. Like all isotopes of iron, this has 26 positively charged protons in the nucleus, accompanied in this case by 30 neutrons. Iron-60 has an extra four neutrons. This makes the nucleus unstable: over time, Iron-60 decays via radioactive cobalt to nickel. Over a period of about 2.6 million years, around half the iron-60 present in a location will decay.

This means that any iron-60 that the Earth got in its hot radioactive beginnings 4.5 billion years ago will be long gone by now. But significant amounts of iron-60 have been detected on the Earth and on the Moon, material which appears to have been created about 2 million years ago. As it happens, there was a mass extinction of marine life around 2.6 million years ago, when we lost about one-third of all marine mammals on Earth. Some scientists have suggested the two were connected. However, there are issues with this – the timing isn't perfect, for example, although all such figures are subject to uncertainty and there might be enough flexibility in the dating for the creation of the iron-60 to coincide with the extinction. Having said that, it's not entirely obvious why the extinctions would have been primarily marine, when ultraviolet that passed through a depleted ozone layer would have more impact on land-based organisms not protected by sea water – but it is a possibility.

If you're worried about friends and family back home, though, don't be. The Milky Way, our galaxy, is a very big place and Earth-influencing supernovae typically only turn up hundreds of millions of years apart. There's certainly no danger from the explosion of Betelgeuse. It will take over 500 years for the light to reach Earth, by which time it will have spread out far too much to have a significant impact.

The ultimate collapse

As we return to watching the end of Betelgeuse, we can also imagine what would happen if the star had been even more massive – starting off with more than 25 times the mass of the Sun. Such stars are rare, but in a galaxy the size of the Milky Way, there are plenty of them. In the case of such a supergiant's collapse, the shockwave from the core bounce would not be enough to blow off much of the outer material, which would keep piling into the intense gravitational field of the neutron core. With the impact of this extra material, even the neutron degeneracy pressure would not be able to stop a further collapse occurring. In such a case, there would be nothing left to stop the process and the result would be a black hole – something we'll be able to visit and learn more about later in the voyage.

Even this isn't the limit – supergiant stars with more than 40 solar masses are predicted to behave differently, although we can't show you this happening live as we aren't aware of a star where this is due to happen soon. In some cases, such massive stars are expected to collapse straight to a black hole, rather than via the intermediate stage of a neutron star, in which case there would be no supernova – the star would simply disappear – but evidence for such events remains limited. Even heavier stars, above the mass of 100 Suns, could produce a special kind of black hole supernova, where the energy production is so great that energy converts into matter, which reverts to energy and so on in an oscillating process until the final collapse occurs. This is known as a pair-collapse supernova, although again, the concept is still purely theoretical with no definitive observational evidence.

As far as Betelgeuse is concerned, we are currently seeing the very start of a process that will last for many thousands of years. The brightness from the radioactive materials will continue for some hundreds of years. By then, the surrounding materials will have gained enough energy to glow in their own right, providing a more dramatic nebula than we would expect from a Sun-sized collapse, and a totally different form of nebula to the cradle for new stars that we saw in the Orion Nebula.

We can't stay around by Betelgeuse long enough to see its remnant take on a clear form, but we have taken a quick trip to one that was made earlier, the evocatively named Crab Nebula.

View 18

Crab Nebula

To see outside the ship, use the QR code or
visit www.interstellartours.co.uk/view18.html

Seen from Earth in the constellation Taurus (although not visible with the naked eye anymore), and sited around 6,500 light years away, the Crab Nebula is thought to be the remains of a type IIP supernova, much like the ultimate fate of Betelgeuse. In this case it's a supernova that was spotted from Earth in the year 1054. The nebula is now around eleven light years across and is composed of a delicate filigree of glowing material with a dark neutron star hidden in its midst. This is how Betelgeuse will likely look a few thousand years from now.

After the initial brightness, and allowing the 600 or so years it takes light to reach Earth from the location of Betelgeuse, observers are likely to start to see something new. Neutron stars might sound like featureless embers, but they can produce one of the most fascinating phenomena in all of the galaxy – a pulsar.

As it happens, the Crab Nebula's neutron star is itself a pulsar, but we are going to visit another such neutron star – one that led to one of the most notorious omissions in the history of the Nobel Prize.

PULSAR

6

Stars don't just twinkle

You are probably familiar with the children's song 'Twinkle, Twinkle, Little Star', one of the best-known in the English language. It was originally a poem by Jane Taylor called *The Star*, published in 1806. This was linked to the familiar tune based on the French traditional melody 'Ah! Vous dirai-je, maman' 32 years later, although the tune is also used familiarly as the Alphabet song, and crops up with a different rhythm as 'Ba, Ba, Black Sheep'. (It's sometimes said that Mozart wrote variations on the theme of 'Twinkle Twinkle', but his piece is actually a variation of 'Ah! Vous dirai-je, maman' as it was written before the familiar words.)

Apart from noting that the comment by the song's narrator 'I know not what you are' is now rather out of date when we know so much about the nature of stars, the key concept from the viewpoint of our interstellar tour is the word 'twinkle'. When we look up at the night sky from Earth, stars don't appear as unchanging bright points of light, but they seem to flicker slightly, varying in intensity.

This distinctive twinkling is not a feature of the light output from the stars. Out here in the vacuum of space, most stars are hard, twinkle-free light sources. The reason for the twinkling we see from the Earth is a combination of two factors. The first is that a star's apparent size is very small – they are pretty much points of light, but sufficiently bright that we can still see them. Secondly, that light can only reach the Earth's surface after it has passed through the atmosphere. As the light does so, it can be slightly displaced from its path, changing direction as it interacts with molecules of gas in the air. But these molecules are not stationary, and so the stars appear to jiggle around slightly, causing the twinkle.

It's said that planets are easily distinguished from stars when observed with the naked eye from Earth because they don't twinkle as much. Although the planets are of comparable apparent brightness, they are a lot closer to Earth, so provide less of a pure point source of light. Theoretically, this would reduce the amount of twinkling – and experienced observers claim to spot this easily. In practice, though, experienced observers also know where the planets are going to be before spotting them, so they may be exaggerating the obviousness of the effect. It exists, but it isn't always easy for the casual observer to spot.

However, the output of some stars varies by more than an atmospheric twinkle. It's almost as if they have a celestial dimmer switch on a timer attached to them, fading and brightening repeatedly. And the first such star that we know that acts in this way was known by the Ancient Egyptians (although they seem not to have noticed its variation in brightness). It's called Algol.

The mystery of Algol

In the early days of professional computer programming, the big-name languages were FORTRAN and COBOL, but a third language had a longer-lasting legacy. This was ALGOL, which would later give rise to the language that dominated for decades, C and its derivatives. ALGOL is a contraction of the words 'Algorithmic Language', but there is no doubt that the name was inspired by our Algol, one of the best-known of the named stars.

This star is formally Beta Persei as it appears in the constellation Perseus and is a relatively near neighbour, located a mere 90 light years from Earth. Algol appears to have been documented by the Ancient Egyptians, but neither they nor the 10th century Arab astronomer Abd al-Rahman al-Sufi, who listed Algol in his star catalogue, mentioned the star's strange trait – however, it does seem to have been imbued with a sense of mystery, as its Arabic name translates as 'the ghoul'. It would not be until the 1660s that Italian astronomer Geminiano Montanari noted that Algol did not stay constant in its brightness but dimmed and brightened over time. It was one of the first observed variable stars.

Later, it was discovered that Algol is not a single star, but a system of three stars orbiting each other. The bigger pair are relatively close to each other, and when the less bright but larger Beta Persei Aa2 passes in front of the smaller, hotter Beta Persei Aa1, the result is a dimming of the overall light output that happens repeatedly every 2.9 days as the stars follow their orbits.

Algol's behaviour is interesting, but it is not the result of variation in the star's output itself. However, a different type of brightness-changing star was later discovered that would put the concept of variable stars on the astronomical map and resulted in the development of a powerful tool for measuring distances in the galaxy. This was a star in the Cepheus constellation that we've already met, although Delta Cephei is nowhere near Gamma Cephei with its attendant planet.

Cepheid variables

One man linked the two stars Algol and Delta Cephei – the English astronomer, John Goodricke. In 1782, he extended Montanari's observations of Algol to show that its variability was periodic – not a random change of brightness, but a regular variation. And just two years later, he found another distinctive variable star with a predictable variation in Delta Cephei, which we now know to be located about 890 light years from Earth.

Thanks to Delta Cephei, we have the name of a whole class of variable stars – Cepheid variables. The mechanism for variation here is entirely different from Algol's. Cepheid variable stars have already consumed the hydrogen in their core and have moved on to later stages of fusion. Around the outside of the star is a layer of helium. When this gas is heated strongly, the helium atoms lose both of their electrons. The 'doubly ionised' helium absorbs a significant amount of light, heating it further.

This results in these outer layers fluffing further outwards. As a result of moving away from the hot core, the atoms cool down, regaining an electron and becoming singly ionised – a form of helium that is more transparent. The outer layer then sinks back inwards, heating up again. The result is a repeated cycle known as pulsation, while the mechanism is called the Eddington valve, after English astrophysicist Arthur Eddington who first proposed it as an explanation for the behaviour of these variable stars.

Because of the varying transparency of that outer layer, the star becomes dimmer and brighter over time. In the case of Delta Cephei, this cycle happens with a period of about 5.5 days. This is all very interesting in itself, but the role of Cepheid variables in measuring distances was why they have become particularly valuable in astronomy, thanks to the work of American astronomer Henrietta Swan Leavitt.

Miss Leavitt's stars

In the early 1900s, when Leavitt was working at the Harvard College Observatory, a site in Cambridge, Massachusetts run by Harvard University, men dominated the astronomical profession, but women were regularly employed as computers. This term originally referred to mathematicians who undertook repeated manual numerical calculations, not the electronic devices familiar to us, which is a usage that dated back at least to the 1600s. But in astronomy, it also encompassed those who made measurements and catalogued data from the photographic plates that by the start of the 20th century had taken over from direct observations by eye through a telescope.

It is under-appreciated just how much photography transformed the science of astronomy. Historically, astronomers observed the skies with their own eyes, initially unaided, and since the 17th century, augmenting their vision using telescopes. Even now, it's not uncommon for astronomers to be portrayed as looking through telescopes. But in practice, as soon as astronomical photography became possible, direct optical astronomy became almost entirely the preserve of amateurs.

The reason was simple. The brightness of a distant object depends on the number of photons that arrive from the object and impinge on the detector – traditionally the human retina. But with a telescope that was motorised to follow the path of the stars (which is to say that can counter the Earth's rotation), the photons from a star could be collected over a long period, building the strength of the image. By the end of the 20th century, this benefit had been extended even further with the replacement of chemical photography by photon detectors – the astronomical equivalent of a digital camera.

This technology not only enhanced light collection even further but meant that astronomers did not have to traipse off to distant telescope locations in the middle of the night. Although early observatories were located wherever was convenient (think of Herschel's observatory in Slough, for example), it was better if they could be built somewhere remote, with less light pollution and disturbance to the air, which degrades the ability to observe. Space telescopes were the best for this, but even with land-based telescopes, electronically collected data meant astronomers could work from their desktops at sociable hours.

More than eyes

Back in Henrietta Leavitt's day, the latest technology available was photographic plates. Clumsy and heavy, as well as requiring operators to suffer through those late-night observing sessions, these were nonetheless a valuable step forward from direct observation by eye. Leavitt realised something that would revolutionise the ability to measure distances to the stars before travel to them was possible.

The distance to relatively close stars had already been measured by using parallax. Hold up a finger at arm's length and close one eye, then alternately open and close your eyes so your view switches repeatedly from one eye to the other. The finger seems to move from side to side against the background. This is known as a parallax effect – and the distance the finger appears to move bears a simple mathematical relationship to the combination of how far the finger is from your eyes, and the distance between your eyes. If you know how far apart your eyes are and how far the finger appears to move, you can calculate how far away it is. A related approach can be used with the nearer stars.

Just switching between eyes with the moving finger doesn't provide a big enough distance to see a variation in the position of the stars in the sky, but astronomers have a way to effectively move their eyes to be 300 million kilometres apart. If you make an observation of the position of a star in the sky, then make the same observation six months later, the two observations are taken from positions 300 million kilometres (186 million miles) apart, as the Earth follows its orbit around the Sun. Time those observations correctly and you get that large movement of the observation position, producing a noticeable shift in the position of the nearer stars against the most distant ones.

However, there comes a point when stars are just too far away for parallax to work. What would make distance measurement easy would be if astronomers had 'standard candles' as they are known in the trade. Imagine we had two stars, each of the same luminosity – they each gave off the same amount of light. Let's say that one of these stars was ten light years away from Earth and the other was twenty light years distant. The apparent brightness of a distant light source depends on the square of the distance away – so the closer star would appear to be four times as bright as the more distant one. If we could measure the distance to the closer star using parallax, we could then make use of these standard candles to discover the distance to the more distant star.

So far, so good. Standard candles are an excellent way to calculate distance. But the method only works if we *know* that the two stars have the same luminosity, something we could only be sure of by visiting each star and measuring the brightness from close by. Enter Miss Leavitt. She spotted something that, given the truth of an untestable assumption, would enable her to identify the luminosity of some stars from a distance. She looked at variability data on a set of Cepheid variables that were all in the same region – the Small Magellanic Cloud, a mini-galaxy and near neighbour to the Milky Way, containing a few hundred million stars.

She found that variable stars of a similar brightness had a similar rate of variation. Assuming these stars were all at a similar distance, which positioning them in the Small Magellanic Cloud (very) roughly makes true, then these types of star of the same luminosity should appear to have the same brightness from Earth. The rate at which the stars dimmed and brightened provided a good predictor of how bright they were. If you picked the right kind of variable stars, you could use them as standard candles to measure many more stellar distances.

In reality, there were two assumptions here. First, that the stars in the Small Magellanic Cloud were indeed at the same distance. We now know it is about 200,000 light years from Earth and around 19,000 light years deep. This would mean if you had two identical stars, one at the near side of the dwarf galaxy and the other at the far side, their brightness would only differ by about 19 per cent – which is not too bad. The other assumption is that the sample of stars used were all the same and all behaved in the same way.

The assumption that all standard candles of a particular type behave the same way remains just that, even to this day. It seems reasonable, but it can only be proved for stars we have seen close up. In fact, the assumption proved dangerous when the approach was stretched to observations further beyond the Milky Way. the first attempt to measure the distance to the nearest large galaxy, the Andromeda galaxy, went wrong as a result. American astronomer Edwin Hubble, using a new, powerful telescope, was able to observe Cepheid variable stars in Andromeda. Using Henrietta Leavitt's technique, he calculated the distance to Andromeda was 900,000 light years.

It was later realised that there were two different types of Cepheid variable star, and Hubble was not comparing like with like. The new distance to the Andromeda galaxy proved to be over 2.5 times larger at 2.5 million light years. Whenever an astronomer uses standard candles, they must be aware of the risk of comparing the stellar equivalent of apples and oranges – but the stars' variability has proved an extremely useful tool.

Henrietta Leavitt was a pioneer in observations of this kind of variability in the output of stars, but it would take another female astronomer to discover a far more dramatic variability that brings us to the neutron star that we have just arrived at, the clumsily titled PSR B1919+21.

Little Green Men

We are now well over 1,000 light years from Earth. Looking out on the neutron star, there is very little to see, just a small, glowing ball far too small to be a conventional star. But when we switch the viewing wall to be able to see radio waves and X-rays, a very different picture emerges.

View 19

Pulsar

To see outside the ship, use the QR code or visit www.interstellartours.co.uk/view19.html

Bear in mind that visible light is just a tiny part of the electro-magnetic spectrum, which consists of the far broader range of radiation that is made up of photons. The spectrum runs in energy from low-energy radio, through microwaves, infrared, visible light, ultraviolet, X-rays and gamma rays. They are all the same phenomenon, merely having photons with different levels of energy. If we think of electromagnetic radiation as waves, it goes from long wavelength, low-frequency radio up to short wavelength, high-frequency gamma rays.

The pulsar, as the kind of neutron star we are now visiting is called, produces a pair of beams in the form of radio waves from its magnetic poles. As we will discover, pulsars rotate extremely quickly: this electromagnetic radiation is the result of a combination of this high-speed rotation and a powerful magnetic field. As the beams of electromagnetic energy rotate, they operate much like a lighthouse's rotating lens produces a sequence of flashes. Should the pulsar's radio beams cross the path of the Earth, they appear there to produce a regular, repeated signal.

The first such pulsar to be spotted from Earth was the one we are now visiting, PSR B1919+21, discovered by the young astronomer Jocelyn Bell (later Jocelyn Bell Burnell) in 1967 at Cambridge University's Mullard Radio Astronomy Observatory. We tend to associate radio astronomy with big dish antennas, like the distinctive relic that is the Jodrell Bank telescope, but many early radio telescopes, like that used by Bell, were arrays of simple wire antennas laid out on frames at ground level – in this case a total of 4,096 aerials forming a 1.6-hectare (four-acre) array. Adding to the distinctly rural feel of the Mullard Observatory, sheep were used to keep the grass down, as there wasn't room to get a lawnmower around between the frames.

The signals from the array were recorded on a long paper roll, showing what Bell would later describe as a 'bit of scruff'. Among the visual representation of background hiss were a sequence of radio blips that came through once every 0.67 seconds, each lasting just 0.04 seconds. This was a spectacular discovery. Although never taken entirely seriously as a possibility, Bell and her supervisor Antony Hewish speculated about the possibility that this regularly repeating radio signal was a message from an intelligent alien lifeform.

This suggestion of an intelligent source for the signal was reflected in the nickname that Bell and Hewish gave it – LGM-1, standing for 'little green men', then a common description of imaginary aliens. No one is quite sure why green was the colour used here. Since medieval times, the 'green man' has been a not-quite-human British folklore figure, perhaps influencing early science fiction, such as Edgar Rice Burroughs' hugely popular Martian novels published from 1912, which included green-skinned aliens.

The far less evocative final name given to the pulsar of PSR B1919+21 reflected not a mistake on the year when it was discovered, but rather the location of the source in the sky based on a 'right ascension' (the angular measurement around the celestial equator) of nineteen hours and nineteen minutes. The speculation that this could be an alien source was soon dismissed when a second such signal was discovered a few months later.

The discoveries at Cambridge were marred by what some regard as an example of sexism on behalf of the Nobel Prize committee of the time. Hewish, and the observatory's head at the time Martin Ryle, were both awarded the Nobel Prize in Physics for 1974 – Ryle for general radio telescope observations and inventions, and Hewish for 'his decisive role in the discovery of pulsars'. But Bell was overlooked, in part because she was only a graduate student, but quite possibly also because of her gender. The greatest astrophysicist working in Cambridge at the time, Fred Hoyle, protested very vocally about Bell's omission from the prize – but to no avail.

The omission has been compared with Rosalind Franklin being left out of the Nobel Prize in Physiology or Medicine for

1962, awarded for discovering the structure of DNA – but the circumstances were very different. Franklin was legitimately rejected because the prize was only available for up to three people (she would have been the fourth) and because she died before the prize was awarded, and it was never awarded posthumously. By contrast, only two received the physics prize in 1974, and Bell Burnell, as she later was, lived on for many years – there was some consolation when she won the 2018 Breakthrough Prize in Fundamental Physics for the discovery of pulsars.

You spin me right round

As it is impossible to get a clear idea of scale out in space without suitable reference points, it's worth emphasising the strange nature of what we can see on the viewing wall. This is no ordinary star. A neutron star, as we've discovered, forms in the aftermath of some types of supernova. It is the remains of the core of a supergiant star, which started off with at least ten times the mass of the Sun and could be up to 25 times that size. After blasting off the outer layers, the remains are in this case around twenty kilometres (12.4 miles) across, but still having one to two times the mass of the Sun.

A pulsar like this is not just a neutron star, but to produce those repeated flashes it is one that has the additional feature of spinning*. And just like its mass, it doesn't do this by halves. The vast majority of bodies in the universe spin. As we saw with the formation of planetary discs, any unevenness in the original gas and dust will result in the matter starting to spin as it contracts,

---

* The vast majority of neutron stars do spin (as do the majority of stars) and are pulsars, but a relatively small number don't have the appropriate level of magnetic field, while old pulsars lose energy and eventually cease to radiate.

because one side will be pulled to the rest more than the other. As a result, not only do planets orbit stars, but planets and stars themselves spin around.

We're familiar with the impact of this rotation on the Earth – it's what gives us day and night. But the Sun is also rotating. This is less obvious, because a star is relatively featureless, and it is too bright to look at directly. But the Sun rotates about every 25 days. Most stars will be spinning around. And when the material compresses to become a neutron star, we get what might be called the ice-skater effect. We've already seen this in the way that the spin develops in a protoplanetary disc, producing planetary orbits and spins, but for the neutron star the effect is raised to a whole new level.

View 20

The ice-skater effect

To see outside the ship, use the QR code or visit www.interstellartours.co.uk/view20.html

As you can see in the video on the viewing wall, an ice skater can dramatically increase the speed of their spin by simply changing the shape of their body. As they move to a more compact position, or pull in their arms, the speed of the spin increases. This is because angular momentum, the oomph of rotation, is conserved – unless energy is transferred elsewhere, such as through friction and air resistance, the angular momentum of a body remains the same. The angular momentum is dependent on a combination of the speed of rotation and the distance the mass is from the axis around which the body is rotating.

When the skater crouches inward, or pulls in their arms, the spin speeds up without any force being applied. Mass is being moved in towards the axis of rotation. As the distance that the mass is away from the axis decreases, the spin speed has to increase to keep the angular momentum the same.

Exactly the same thing happens to the remnant of the much bigger original star that is the neutron star – but on a far greater scale. As long as the original star is rotating (which it almost certainly will have been), the neutron remnant will rotate far faster. Remember this is the equivalent of the Sun dropping in diameter from 1.4 million kilometres (870,000 miles) to twenty kilometres (12.4 miles) across – that's pulling in a whole lot more than just a skater's arms.

As a result, the pulsar we're located next to is able to spin round not in days, nor in hours, but every 1.34 seconds. (The pulses it produces are detected every 0.67 seconds, but they are the result of receiving beams emitted from both poles.) And this first discovered example is a laggard by pulsar standards. Some rotate in just a few thousandths of a second. The record holder for some decades has been PSR J1748−2446ad, which rotates at over 716 times a second. This is astonishingly fast. It means that its surface is moving at around 24 per cent of the speed of light – over 70,000 kilometres (43,500 miles) per second.

Not a great neighbourhood

After leaving the first pulsar to be found, we're now en route to another historical milestone location in the detection of pulsars – the first pulsar discovered that was known to have planets of its own. This is PSR B1257+12, which, unusually for a pulsar, also has the more manageable name of Lich. The word is a variant spelling of 'lych', most commonly found in the term 'lych-gate', which refers to a covered gateway often found at the entrance to a churchyard, which was built to provide shelter for a coffin on the way to its final resting place. A lich was an old word for a corpse, although given the names of the other planets in the system, the intention was probably the more modern fictional extension of the term to apply to an undead creature in a fantasy setting.

PSR B1257+12 is a faster-spinning pulsar than the first to be discovered, rotating in around 6.22 milliseconds. That's about 160 rotations a second – for comparison, the fastest rotation rate of old audio CDs was about eight rotations per second. Lich's spin rate was discovered to be less consistent than that of a normal pulsar, which was the first indication that this stellar relic still had planets. Indeed, two of its planets were the first ever exoplanets to be confirmed – Draugr and Poltergeist, names that we've already encountered, related to mythical Norse undead creatures and violent ghosts.

View 21

Lich

To see outside the ship, use the QR code or
visit www.interstellartours.co.uk/view21.html

The third planet was later given the name Phobetor. Once again, astronomers were going for a name with spooky connotations. Given the similarity to Mars' moon Phobos, the Greek god of fear, it's not surprising that Phobetor (translating as 'Frightener') was the name of a beast-like creature who was one of the sons of Somnus (another son was the better-known Morpheus) in Ovid's poem *Metamorphoses*. It's alongside Phobetor that we've emerged from hyperspace, giving us a clear view of the three planets and the dramatic form of the pulsar. To do this, we are showing its emissions with a wide range of electromagnetic radiation visible. The star itself is intensely glowing with high-energy light – its surface temperature is over 28,000°C (50,400°F), more than five times the surface temperature of the Sun.

Planets are extremely unusual around pulsars, because the dramatic explosion of a supernova would usually destroy any planets that hadn't already been absorbed by the supergiant stage of the star. In this case, the three planets are thought to have formed after the explosion from the remnant matter dispersed by the supernova that occurred when two white dwarfs merged over a billion years ago. The shattered remains were pulled together by gravity, allowing a new extrasolar disc to form. As is not entirely surprisingly for planets orbiting such an extreme star, there is no possible Goldilocks zone and no sign of life.

If you were to get near to Lich (or any other neutron star for that matter), the gravitational pull would be intense. This is not only because this is a body with more mass than the Sun – so with a far stronger gravitational field than the Earth – but also because its compact form means that you can get much closer to the centre than you normally could to such a massive body. As the gravitational pull increases at the square of the rate of the decrease of the distance from the body's centre, this makes for extremes of force that can produce dramatic effects. But rather than study them here alongside a neutron star, we're going to take a step further to investigate a stellar remnant that is even more extreme. The most extreme type of object in the galaxy.

It's time to take a trip to a black hole.

BLACK HOLE

7

A dark star

When you first came on board the ship, we took a look at what was necessary to get off the Earth and into space. We discovered that the Earth has an escape velocity of 11.2 kilometres (seven miles) per second – if you throw something upwards towards space at this speed or greater, then it will continue out into space. But if you throw it slower, it won't make it all the way, falling back to Earth. (We also saw how rockets can travel at much slower than escape velocity, because they are under power, rather than just thrown.)

The idea of escape velocity followed directly from Isaac Newton's work on gravity in the 17th century, meaning that the concept was understood by scientists long before anything was ever sent from Earth into space. One individual who certainly knew about it was the English astronomer John Michell. Despite being born as early as 1724, he was the first person to come up with the idea of what would later be referred to as a black hole.

For many centuries prior to Michell's work, the speed of light had been a mystery. Galileo Galilee had tried to measure it by flashing a light at a distant observer, who would then flash a light back, timing how long the process took. But it didn't work. Not only did Galileo not have a precise enough timepiece, relying mostly on his pulse rate, he discovered that the only time interval he was measuring was his own reaction time. As far back as ancient times, some individuals had thought that light travelled infinitely fast, arriving the moment it left. Others thought that, though its speed was clearly high, it was not infinite. As Galileo ruefully put it: 'If not instantaneous, it is extraordinarily rapid.' But the existence of a finite speed was settled (unintentionally) by the Danish astronomer Ole Rømer in 1676.

Rømer was observing the moons of Jupiter through a telescope and found that the timing of their orbits around the planet seemed to vary, depending on the relative position of the Earth and Jupiter on their paths around the Sun. It was realised that this variation was the result of the time it took the light to get from Jupiter to Earth. Although Rømer never made the calculation himself, somewhat later, his measurements were used to provide a first estimate of the speed of light, which worked out at about 220,000 kilometres (136,700 miles) per second.

Measurements of the speed of light were refined over the years, although it is now fixed at an exact value*, and although the rate had not been totally settled by the time Michell was thinking about escape velocity, the scale of the speed was understood. Michell was not a professional astronomer – but then, professional scientists were a rarity in those days. He had been an academic, rising to become Woodwardian Professor of Geology at Cambridge University. But by 1767, he

---

* The speed of light in a vacuum is now exactly 299,792,458 metres per second, because, since 1983, the metre has been defined as the 1/299,792,458th of the distance light travels in a second. That makes light speed around 1,080,000,000 kilometres per hour or 671,000,000 miles per hour.

had abandoned academia to take over as rector of St Michael's Church in Thornhill, West Yorkshire.

Perhaps the less-demanding role as a clergyman in a large village gave Michell more time to think. Certainly, he made Thornhill something of a centre for scientific discourse, holding an 18th-century equivalent of a modern scientific conference in 1771 that saw the discoverer of oxygen Joseph Priestley, photo-synthesis pioneer Jan Ingenhousz, civil engineer John Smeaton and political and scientific polymath Benjamin Franklin head off to the wilds of Yorkshire to discuss scientific matters.

By 1783, Michell was ready to present a paper to the Royal Society in London on his concept of dark stars. Given a fixed speed for light, Michell realised something surprising. If the escape velocity of a star was high enough, it could be greater than the speed of light. In that case, he presumed the star would appear black. No light could ever escape from it because it wasn't going fast enough. It would form what we would now call a black hole. To come to this conclusion, Michell assumed that corpuscles of light escaping from the star would be slowed down by the pull of gravity, just as a thrown ball was. These 'corpuscles' are what we would now call particles.

Although the wave theory of light already dominated in Europe by Michell's time, like many British scientists, he sub-scribed to Newton's theories that light was made of particles. This idea was pretty much dead by the 19th century, only to be proved correct thanks to Einstein's Nobel Prize-winning work on the photoelectric effect that formed one of the foundations of quantum theory.

The way that Michell thought of such a dark star could no longer be applied once the true nature of light was understood. When the Scottish physicist James Clerk Maxwell showed in the mid-19th century that light as an electromagnetic wave had to travel at a fixed speed in any particular substance, it was no longer possible to imagine that light would slow down like a projectile as it left a star, fighting against gravity. But that still left the question of what the implications were for a body that had an escape velocity that was greater than the speed of light. To resolve that would take Einstein's greatest work – the general theory of relativity.

Gravity becomes relative

Einstein's masterpiece explains how gravity works. Where Newton had successfully shown how to calculate the force of gravity, he explicitly claimed that he 'framed no hypotheses on how it worked'. (He did, in fact, have a theory, but he put that in a separate publication.) In 1915, Einstein showed that gravity is the result of matter having a direct influence on the space and time around it. Anything with mass (or energy) warps spacetime. The more massive the object is, the more it distorts its spacetime environment.

This is why, for example, the Earth orbits the Sun. As far as the Earth is concerned, it is happily travelling in a straight line, not experiencing any force. But the mass of the Sun warps spacetime sufficiently to cause the Earth to go around it in an orbit. This aspect of gravitation is often likened to placing a bowling ball on a flat sheet of rubber. If we imagine the Earth's straight-line path as a line drawn on the rubber, when we place a bowling ball on the sheet, near to the Earth's path. The ball will produce a dent in the rubber sheet that will result in that straight line becoming a curve.

This doesn't explain why things fall under the force of gravity, because when we let go of them, they are not moving, so we don't have a straight-line trajectory to be warped with space. This is where we have to get our heads around the concept of spacetime – the combination of space and time, with time seen as a kind of fourth dimension. According to the general theory, matter doesn't just warp space, but warps spacetime. Mindbogglingly, it's actually a twist in the time dimension that results in the acceleration of a dropped item towards the Earth.

Unlike the special theory of relativity, the mathematics involved to develop the general theory was tortuous – in fact, Einstein struggled with it and had to get help, particularly after making a few mistakes that meant he was nearly pipped to the post in publishing a correct theory by German mathematician David Hilbert, who had not had the original ideas but was a better mathematician.

Einstein finally produced his correct formulation towards the end of 1915. His theory describes the interaction between a body with mass and spacetime in a complex set of equations that even now can't be solved in a general way. In fact, Einstein's initial assessment was that it would be unlikely for solutions to ever be developed for his equations. However, this was one of the shortest-lived incorrect predictions in the history of science.

Just months after Einstein published his theory, the German mathematician Karl Schwarzschild was playing with the implications of these equations in the unlikely setting of the First World War frontline. Schwarzschild was too old to be called up to fight for his country, but despite being past his 40th birthday, he volunteered for the army. For most people, the front would have been the last place to indulge in deep thinking – but perhaps Schwarzschild found physics a useful distraction from the horrors of war.

By focussing on a very specific and small-scale application, Schwarzschild was able to solve a reduced version of the equations. Physicists have a habit of simplifying things as much as possible to keep their mathematics relatively accessible. This is sometimes portrayed by imagining a physicist thinking about solving a set of equations for the dynamics of a moving cow. The first line of the work is: 'Let us assume that the cow is spherical.' Schwarzschild decided to work with a gravitational equivalent of a spherical cow.

The cosmological spherical cow

Schwarzschild imagined a body that was totally spherical and had no electrical charge. It did not rotate, and it was in a universe where a specific constant, known as the cosmological constant, that was part of Einstein's equations set to zero, which simplifies things by taking out a major chunk of the messy mathematics. This was not Schwarzschild being ahead of his time and somehow envisaging the nature of a black hole (in the modern sense of a black hole, rather than Michell's) before anyone had heard of one – he was really just playing around with a simple object that made the equations easier to handle. What might now be called a 'toy model'. At the time, no one would have suggested that such a thing existed in reality.

However, Schwarzschild's solution had some mind-bending implications, should anything real ever be discovered that did exist in this form. His solution to Einstein's equations described a specific type of body that was so dense that it produced an extreme warp in spacetime. A distortion that was so severe that any light escaping from that spherical, uncharged, non-rotating object would be bent back into its body. The light could never emerge.

This was a rebirth of Michell's broad idea, but one that worked with the reality of light's nature. At the time, this was not thought to concern anything real. As we have seen, it was just a way to play around with the implications of Einstein's equations. Einstein himself was entirely convinced that such an object could never be found. He considered it impossible to imagine a physical process that could compress matter as much as would be required to produce this effect.

Of itself, a black hole could be any size – it's 'just' a case of compressing matter sufficiently and a black hole becomes inevitable. If a comic book villain came along with a super weapon that could compress the Earth until it was smaller than around 0.9 centimetres (0.35 inches) in diameter, for example, it would become a (tiny) black hole. As Einstein suggested, though, there is nothing that we know that could do this to the Earth. Our home planet is safe from this particular type of villain.

However, as the understanding of the life and death processes of stars was refined, it was realised that there might be a way for such an event to happen out in the universe if there were a big enough star. The reality is that the only process we know of that could produce that much compression is the gravitational attraction of a massive body, such as a star that is considerably bigger than the Sun and that has lost its ability to sustain itself*.

The familiar term 'black hole' was not introduced until the 1960s, long after Schwarzschild's work. Its first use has often been attributed to the American physicist John Wheeler, who certainly popularised it in 1967, but it seems that it was not Wheeler who originated it. The first recorded use of the term was at a January 1964 American Association for the Advancement of

---

* Strictly speaking, there is another mechanism by which black holes could have formed in the early history of the universe, when the first matter occupied a relatively small volume of space. 'Primordial black holes' have been hypothesised, formed by localised high densities of matter in the high-energy state of the early universe. These could have been a lot smaller than black holes that are formed by stellar collapse. However, there is, as yet, no evidence that such black holes exist.

Science meeting. Unfortunately, it was not recorded who used the term, but the article mentioning it in *Science News-Letter* was written by Ann Ewing. It has been suggested the term was first used a couple of years prior to this by the American physicist Robert Dicke, but this is not well-documented and does not seem to have been the origin of the phrase's wider usage.

Welcome to Cygnus X-1

Before we find out more about black holes, let's take a look at one. We've just arrived in the vicinity of Cygnus X-1. This was the first black hole to be detected – though initially it was just seen as an intense source of X-rays from space (hence the 'X-1' in its name). From the Earth, it is located in the constellation Cygnus and is the most distant location that we've visited so far – over 6,100 light years from home.

View 22

Cygnus X-1

To see outside the ship, use the QR code or
visit www.interstellartours.co.uk/view22.html

Interestingly, Cygnus X-1 could not have been detected much earlier than it was, which was in 1964. This is because X-rays from space are not able to penetrate the Earth's atmosphere.

(This is just as well, really.) As a result, it was not until relatively low-flying rockets carrying Geiger counters were used to scan space that this black hole was discovered.

Looking out on what we are claiming to be a black hole seems to suggest that the name is anything but accurate. The view that we can see is certainly not of an entirely black object, seen against the slightly less-black background of space. Part of the reason for this is the big bright blob to the left of your view. This is not the black hole itself, it's a separate, blue supergiant star with the uninspiring name of HDE 226868. This star and the black hole form part of a binary system, orbiting each other at about half the distance that Mercury is from the Sun.

Speculation alert

Given the amount of coverage they get in the news media and sci-fi movies, it might be something of a surprise to learn that the very existence of black holes remained speculative well into the 21st century. There were certainly bodies out there in space that behaved the way we would have expected a black hole to behave. But the nature of a black hole is purely theoretical, and even with the technology to go and visit them, we can't know for certain what is actually going on inside what appears to be a black hole. In 2023, we were just starting to get direct images of a black hole, but, as we will discover, the reality of their nature can never be directly observed.

A shifty star

It was the colour of HDE 226868 that gave the first indication of the existence of the black hole lurking in the system. Way back in 1971, a radio source had been identified with this star and the following year, Australian astronomer Louise Webster and English astronomer Paul Murdin at the Royal Greenwich Observatory, along with American–Canadian astronomer Charles Bolton at the David Dunlap Observatory in Toronto, spotted a Doppler shift in the star's output.

You have probably heard the change in pitch that occurs when a vehicle with a siren drives past you. As the sound source passes by, the noise shifts from a higher to a lower pitch. This is a Doppler shift, named after Austrian physicist Christian Doppler. Technically, the pitch of a sound is described as its frequency – which simply refers to the number of times the sound wave goes through a cycle in a second. In the case of sound, such a wave involves air getting squashed or compressed together, then relaxing, in a repeated pattern.

Doppler reasoned that when whatever's emitting the wave comes towards us, that movement will mean that in the time between emitting one peak of compression of the air and the next one, the source will have moved forward, meaning the gaps between the highest and lowest moments of compression will be reduced. The sound's frequency will be higher than it would be if the source wasn't moving. Similarly, as the sound emitter moves away, the gaps between the peak compressions will increase and the waves will become stretched out, so the peaks and troughs are further apart. This means that the peaks come less quickly – the frequency drops.

Doppler came up with this idea in 1842, although it would take Dutch meteorologist Christophe Ballot to demonstrate the effect in hilarious fashion in 1845. Ballot got two groups of trumpeters together. One group was positioned in a wagon behind a steam engine, the other on the platform of the station. All played the same note – but as the train rattled past at speed, the notes produced by the mobile trumpeters went from too high to too low.

A very similar thing happens with light. In fact, Doppler was specifically interested in the colour of light produced by binary stars. When a star is moving towards an observer, if we think of light as a wave, the wave gets squashed up*. Because of the wave being squashed up, the light coming towards us increases in frequency, just like those trumpets. An increase in frequency moves the light up the colour spectrum – it is described as being blue shifted.

These days, we often prefer to think of light as a stream of photons rather than as waves. The equivalent of the frequency of the light is the energy of the photons. It's easy enough to see that if a source is moving towards us, we would expect the photons to have higher energy (just as a punch will have more energy if someone runs towards you at the same time as throwing the punch). And higher-energy visible light is more towards the blue end of the spectrum.

Similarly, as a light source moves away, the frequency of the waves will reduce – the energy of the photons will drop as if someone is pulling a punch. The result is that the light is red shifted, moving down the spectrum towards the red. It's ironic that Doppler came up with his effect to explain a shift in colour of the light of stars, but red shift or blue shift is usually explained by referring to the Doppler shift of sound. (Doppler actually got the effect a little wrong as far as stars were concerned, and his ideas had to be sorted out by the French physicist Armand Fizeau.)

What the astronomers checking out HDE 226868 discovered was that its colouration underwent a Doppler shift. Every 5.6 days it went through a cycle of blue and then red shifting. It

---

* In the case of light, the wave is a side-to-side oscillation compared to its direction of motion, as opposed to the compression and rarefaction of a sound wave.

was doing exactly what Doppler intended his effect to indicate. The star was part of a binary system with another star. From the data, the astronomers were able to get an approximate idea of the mass of the invisible other star. And it was big. Far too big. For reasons we will discover shortly, the astronomers were able to excitedly conclude in their paper, 'it is inevitable that we should also speculate that it might be a black hole'.

Passing the neutron star limit

The bit on the right of the view we can see, which *is* the black hole, has a tiny black core, but the vast majority of what we can see is in the form of a fiery spiral. We will explore more of the detail of what's going on in and around the Cygnus X-1 black hole in a moment, but let's briefly take a step back and see how the theory behind them developed. We have already visited neutron stars – themselves weird physical structures. And for stellar remnants up to a certain mass, a neutron star is the likely outcome. But if the star is too big, there's a problem.

Neutron stars form, as we discovered earlier, when the immense gravitational force making them collapse is balanced out by neutron degeneracy pressure. We saw electron degeneracy stopping the core collapse of the Sun earlier on, and similarly in a neutron star it's the equivalent process with neutrons that counters the collapse, although it allows a much closer packing before it can resist the pull of gravity.

However, there is a limit to this resistance. If the gravitational pull is too high, even neutron degeneracy pressure can't hold it back. A white dwarf remnant will remain stable if it is less than 1.4 times the mass of the Sun. A neutron star can resist collapse up to something between 2.2 and 2.9 times the mass of the Sun (this would correspond to an original star, pre-supernova of perhaps twenty times the mass of the Sun or more) – but beyond that, the force of gravity is greater even than the neutron degeneracy pressure can resist. There is nothing left that we know of that will stop it from collapsing. If a star exceeds the clumsily named Tolman–Oppenheimer–Volkoff limit, it becomes a black hole.

Such a collapse can sometimes produce a supernova, as is usually the case when a neutron star forms, but in the case of Cygnus X-1, this seems unlikely. A supernova is so disruptive that the black hole would be very unlikely to stay in its close relationship with HDE 226868 after such an explosion. What is more likely is that the core simply collapsed with only a relatively small blowing out of the rest of the mass of the original star that became Cygnus X-1. Some of its material would have been picked up by HDE 226868 with much of the rest being dispersed out of the system.

Total collapse

In theory, then, a black hole is a star that has totally collapsed to an infinitesimally small point. This leads to various aspects of the star, notably its density, in theory becoming infinite. Generally speaking, when any physical theory predicts an infinite outcome, it reflects a failing in a theory or a theory that has exceeded some limit and needs revising. The chances are that in the physical world, as opposed to the mathematical world of theory, something else will happen. This is why the black hole remains a speculative concept in the fine detail of what's going on in there. We don't know of anything that will stop its total collapse – but then, we don't know what we don't know. There may be something else that does prevent such a collapse. Or something totally different may happen.

All observations can tell us is that there are objects out there in the galaxy that behave very like black holes. But exactly how they are structured, how they work internally, is impossible to investigate however closely we examine them – and it always will be. Even if there were a 'not quite all the way black hole' – which may be what we are observing here at Cygnus X-1 – it will still prevent us from discovering what has actually happened when it collapsed.

Remember that a black hole occurs when the gravitational pull becomes so strong that even light can't escape. If light can't get out, neither can anything else. Once you get past that point of no return, known as the black hole's event horizon, there is no escape. All we can see is the event horizon, not the black hole itself. So, we have no way of probing the interior and getting information about what's going on inside a black hole. And it would be perfectly possible to observe what we can see here near the black hole without it having collapsed inside that horizon to an infinitely small point.

If your only experience of black holes is from science fiction, you may have some misconceptions about the way that they behave. They have often been portrayed as black, invisible menaces. We might expect them to act like cosmic garbage disposal units, sucking in anything and everything that had been in their former solar system. You may also have seen them portrayed as gateways to travel to another universe or an alternate dimension. And although each of these ideas has been influenced by the truth, the reality always requires significantly more nuance than fiction allows.

So, yes, a black hole is black in the sense that nothing emerges from it, including light. That makes it more black than anything else, as a normal object that is apparently black gives off some types of electromagnetic radiation, despite absorbing most of the visible light. Even the blackness of space contains stray photons from distant stars. Yet, as we can see with Cygnus X-1, it's perfectly obvious where the black hole is, because it is surrounded by that glowing swirl of material.

Although the black hole itself is not a source of light, if there's anything close enough to be gravitationally attracted to it, which isn't in a stable orbit, it will be accelerated to a very high speed before it passes through the event horizon and disappears for ever. As this happens, this accelerating material will be heated up to the extent that it glows brightly – producing, for example, the strong X-ray signal that made Cygnus X-1 stand out in the first place. The temperatures are so high in a black hole's 'accretion disc' – potentially reaching millions of degrees – that matter is converted to energy more efficiently than it is in a star. Vast amounts of electromagnetic radiation, much of it in the form of X-rays, pours out.

As for a black hole's reputation of sucking in everything around it – this is no more or less true of a black hole than is the case with a conventional star. The black hole has the same gravitational pull as a star with the equivalent mass – in the case of Cygnus X-1, about fifteen times the mass of the Sun. If we took the *Endurance* into orbit around Cygnus X-1 at an appropriate distance, we would be just as safe as we would be orbiting a similar-sized star. However, as with a neutron star but even more so, matter can get much closer to the black hole than it ever can to a normal star – but get close enough and the gravitational effects that are experienced begin to exceed anything that is normal.

This perfectly normal attraction is why you can see material from HDE 226868 streaming in towards Cygnus X-1. The stars are very close together and there is sufficient gravitational pull for the black hole to attract loosely attached material from the star, gradually growing the size of the black hole. But material from the black hole can't be pulled out by the companion star.

As for black holes acting as travel gateways, this is, as far as we can tell, pure science fiction. We certainly have no intention of taking the *Endurance* into a black hole to see where we end up. The whole point of a black hole is that there is no escape. Once you pass through the event horizon, there is no way out. Cosmologists love playing around with hypotheticals, and if you happened to have to hand a totally speculative object called a white hole, then you could in principle combine it with a black hole to make a gateway. But there is no evidence for the existence of white holes, which, as we have seen when considering wormholes are in effect anti-black holes which nothing can enter.

A case of spaghettification

Before long we are going to send a probe towards Cygnus X-1. But before we do that, let's think a little about the experience of approaching a black hole. As we've heard, the point (or, rather, the sphere) of no return is called the event horizon – that's about 44 kilometres (27.3 miles) from the centre of this particular black hole. The distance is known as the Schwarzschild radius, as it was first calculated for Karl Schwarzschild's simplified solution of Einstein's equations. The central point of the phenomenon is referred to as a singularity – but this is a reference to the quite possibly fictional notion that the black hole has totally collapsed and has zero radius. Even so, it's still a useful point of reference.

That event horizon is a big deal as far as light is concerned, but it is not any kind of physical barrier or membrane that has to be passed through to reach the interior. There would be no specific experience for someone who was passing through it – in fact, with a big enough black hole you could get close enough to pass through the event horizon before experiencing any extreme phenomena. But once you had, there would be no coming back. With a black hole the size of Cygnus X-1, though, things would have become distinctly distressing before passing through this sphere of no return.

As you get closer to an object with mass, its gravitational attraction gets greater. The force increases with the inverse square of the distance away from its centre. So, if you halve the distance, the gravitational force increases by a factor of four. (This is standard Newtonian stuff.) In this respect, a black hole is no different from any other massive body like the Earth or the Sun. But bear in mind that we have in Cygnus X-1 a body that has a mass of around twenty times that of the Sun, but which is effectively non-existent in size. You can get as close as you wish to.

This doesn't just mean that the force accelerating an approaching body towards the black hole becomes immense as it gets closer. It also means that over a relatively short distance the force gets considerably greater. And this can produce dire results.

Imagine for a moment you were approaching Cygnus X-1 feet first in a spacesuit. You would get faster and faster as you approached. But, also, you would increasingly feel a stronger force pulling your feet than was pulling your head, because your feet were a little closer. It would be like being on that medieval instrument of torture the rack, but with vastly more force involved. Before long, you would begin to stretch. And that process would become more and more extreme until you ended up as a long, thin cylinder of material, a process that cosmologists have called spaghettification.

Just to get a feel for the scale of the force that is stretching you, let's throw some numbers into Newton's equation for the force of gravity, which is:

$$G \times m_1 \times m_2 / r^2$$

G is just a constant amount (6.6743×10^{-11} if you want to know, assuming we're dealing with scientific units), m_1 and m_2 are the masses of you and Cygnus X-1 respectively, and r is the distance you are away from the centre of the black hole. Plugging in the relevant values, the force you would experience between your head and your feet if you were about ten kilometres (6.2 miles) from the singularity would be around 1.33×10^6 newtons.

The newton (with a small n) is the scientific unit of force. From the calculation above, the force you would feel pulling you apart would be around 1.33 million newtons. To put this into context, imagine the kind of force it takes to get a good-sized passenger train moving. The force exerted by the black hole at the point mentioned above is about the same. You can imagine what would happen if one end of a person was attached to something solid and the other end was attached to a train engine that started to accelerate away. And as you get closer and closer, that stretching force would increase more and more quickly. Survival would not be an option.

Don't forget relativity

We've now launched a probe towards the black hole. This is taking a route out of the plane of the whirling mass of material around Cygnus X-1 to avoid being wiped out by the high-energy particles that are being swept around. If there were humans on board, they would soon be killed due to the level of radiation being given off by the accelerating matter that is plunging into the black hole, but thankfully, our probe is unmanned and protected by an impressive amount of shielding. Even so, it would not last long under normal circumstances.

If there were people inside a ship following the same course, who were protected against the assault of the radiation, they would experience time flowing in the normal way and would pass through the event horizon at the time you would expect for a craft moving at the speed this one is. This doesn't mean, though, that approaching the black hole would look the same to them as making an approach to, say, the Earth from space. This is because of the way that the black hole's gravitational field distorts the space and time around it.

Where an approaching planet or moon gradually gets bigger in our view, the black hole will appear from the probe's viewpoint to balloon outwards, because the passage of any light around it would be distorted, making the central blackness get bigger and bigger. Astronomer Becky Smethurst has pointed out that if you were approaching the Moon and were ten times its diameter away from the surface, the Moon would look the size of your fist at arm's length. By contrast, if you were ten times the size of the black hole's event horizon away from its event horizon, the black hole would fill your entire field of view.

As the probe gets closer still, that warping of light will stretch back around it, until it seems that the black hole is visible in every direction, with just a tiny spot directly behind the probe, where the light that is coming from behind the black hole is bent around and comes towards you. Finally, as the probe passes through the event horizon ... who knows? All we know for sure is that now all roads lead to the theoretical singularity in space and time – if that truly exists. Whichever direction you tried to move, that is the way you would be heading.

Yet, from our viewpoint, orbiting Cygnus X-1 at this distance, we can see something very different. Two relativity effects mean the probe's time is slowed down from our viewpoint.

Firstly, as the probe gets nearer to the black hole, that rapidly increasing gravitational pull is accelerating the probe to high speeds. And secondly, the probe will experience a stronger and stronger gravitational field. Taking that acceleration first, as we saw in the first chapter, according to Einstein's special theory of relativity, from our viewpoint, time on the probe would slow down. This is not simply a distortion. We are familiar with something like perspective that makes things that are further away look smaller than they really are. But they don't actually shrink as they move into the distance. Here, as far as we are concerned, time genuinely is running slower on the probe.

Our view of the probe in this instance reflects one of the first experiments on Earth to confirm the special theory of relativity. The Earth experiences a constant flow of particles from deep space known as cosmic rays. These 'rays' arrive at high speed and collide with atmospheric molecules, producing showers of newly created, high-speed particles. Some of these new particles called muons have very short lifespans before they decay into other forms, but they are travelling fast enough that time runs slowly for them as seen from Earth, and they survive significantly longer than they otherwise would.

If we could see a clock on the probe from our vantage point, it would run slower and slower as it got closer to the black hole. The mathematics to prove this is high-school level – we won't go through it right now, but you have been provided with an appendix at the back of the book that takes you through it, if you'd like to give it a try. Our imaginary crew on a manned version of the probe would not see their clock slowing down. This is why it's called relativity – the flow of time (among other things) depends on how you are moving relative to the clock.

Simultaneously, time would also slow on the probe from our viewpoint because of the gravitational field of the black hole. The general theory of relativity (where the mathematics is anything but school level, as we've seen, challenging even Einstein in his day), shows that the stronger the gravitational field, the more the time will slow down on the probe from our viewpoint.

Another way these effects were demonstrated on Earth back in the 20th century was crucial to one of the most important satellite systems of the day. Satellites move with respect to the Earth's surface, so time runs slower on them as seen from Earth. This special relativity effect was first demonstrated in 1970 by flying atomic clocks around the world, measuring the loss in time when they returned. (The researchers couldn't afford a private plane – they booked a pair of seats for a Mr Clock to accompany them on their flights.) The gravitational effect of slowing time was measured by sending gamma rays down a tower and measuring the blue shift of the light. This showed that, on a satellite, time should run quicker than on the Earth's surface as it experiences a less strong gravitational field. The two results are contradictory, yet both are real: the combined effect of the two had to be dealt with to enable the late 20th-century satellite navigation system Global Positioning System (GPS) to work.

The GPS system was based on a series of highly accurate clocks, positioned in satellites that orbited at around 20,200 kilometres (12,600 miles) above the Earth's surface. By picking up the time from a number of satellites, hence measuring distances from the time the radio signal took to arrive, a receiver on Earth could work out its position. But the writers of the GPS software had to allow for both these effects of relativity. The movement of the satellites made the clocks run slow, but because the satellites were further from the Earth than a clock on the ground, they ran fast. This general relativity effect was the larger of the two – if the system had not been corrected to deal with this, GPS positions would drift out of alignment with reality by up to ten kilometres (6.2 miles) a day.

As we observe the probe nearing the black hole's event horizon, our measurements show that it is moving slower and slower. From the viewpoint of the probe, although it is still accelerating. But because its time is passing slower and slower from our viewpoint, we can see it getting slower and slower. We will never see it pass through the event horizon, even though, as far as someone on the probe is concerned, it will pass through unhindered.

At the same time, our view of the probe will become increasingly distorted as it dims and it changes colour. As physicist Sabine Hossenfelder has pointed out, when the probe nears the black hole it won't just get dimmer, the light you receive from it will also get shifted to the red. Most of the changes in appearance are due to the light from it being bent more away from a straight line by the black hole's gravitational field, but tidal effects would begin to stretch the ship as well – however, in the time we have to observe it, we won't see any significant spaghettification.

It's almost impossible to think of the singularity (should it exist) as anything other than a point in space – the centre of the black hole. But bear in mind that massive bodies do not just warp space, they warp space and time together. The inevitability of a singularity is more that it is a point in time. Once something has passed through the event horizon it should arrive at the singularity at that point in time. Nothing can stop it from doing so.

Variations on a black hole theme

What we've thought about so far as a black hole is what's often called a Schwarzschild black hole – that is a black hole that does not rotate, which was one of the assumptions Schwarzschild originally made to make the solution of the equations of the general theory of relativity less complex than they otherwise would have been.

However, as we've already seen, this is a special case. It is pretty much the norm for real stars to spin around. And if the star from which a black hole formed was spinning, we would expect the black hole to spin too. Like a neutron star, in fact, it should be spinning at extremely high speed, although in the case of the black hole, it is difficult to say exactly what is spinning. Bear in mind that the event horizon is not a physical thing – if a black hole with no radius really does form as theory predicts, arguably it should have an infinite spin speed, which theory predicts might make its singularity become a ring, rather than a point.

By 1963, New Zealand-born physicist Roy Kerr was able to take Schwarzschild's work further and provided a solution to the equations of the general theory of relativity for black holes that were rotating. One of the implications of the theory when applied to rotating bodies is a feature called frame dragging. When a massive body rotates, it drags around nearby space and time with it, a bit like rapidly twisting a spoon in a jar of honey will cause the honey to start to follow it around, and this is known as frame dragging. In the case of a black hole, Kerr found that this frame dragging had a couple of consequences.

The first is that nearby matter must follow around with the twisting spacetime. When space itself is manipulated, there is no issue with the light speed barrier – light itself will be dragged around – and if matter managed to keep still, it would be travelling faster than light. The spin would also have an influence on the size of the event horizon, which would contract. This is because the force of frame dragging would in part push outward, enabling light and matter to still escape that would otherwise have passed the event horizon of a Schwarzschild black hole.

Whatever is actually spinning, we know that something is, because that's why we can see Cygnus X-1 with that distinctive disc of rotating matter around it. We'll filter out the rest for the moment and take a look at the key elements of the black hole itself and its surrounding, high-speed rotating disc of matter.

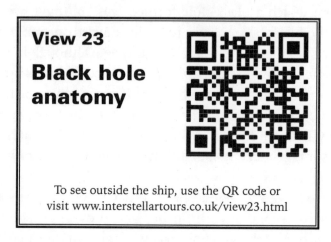

To see outside the ship, use the QR code or visit www.interstellartours.co.uk/view23.html

Without the inflowing matter, we can see on the viewing wall a strange, asymmetric appearance that changes as we move the ship to see the black hole from above, rather than side-on. The hat-like asymmetry seems particularly odd in what must be one of the most symmetrical bodies in the universe. After all, black holes have no surface features – they have no surface. If we could take away the distortion from the black hole bending spacetime, there would be just the black centre and the disc of glowing material around it.

The top-side ring is actually a continuation of the disc of glowing material around the black hole, but what is seen above the hole is actually the back side of the disc, its light's path warped by the hole's gravitational field. The bottom-half ring is light heading out of the underside of the disc that again is being warped – the asymmetry comes from the relative positioning of the different parts of the accretion disc and the viewer.

The apparent different strata of the disc, making it look almost like the rings of Saturn had it not been distorted, is the

Doppler effect at work again as it is blue shifted when it moves towards the viewer, producing the higher-energy, more-yellow light, and it is red shifted as it moves away, producing the lower-energy, redder light.

Hawking's legacy

We can see from Cygnus X-1 that a black hole with stuff around it produces a fair amount of radiation – but what would happen if we came upon an isolated black hole? They are relatively rare, because black holes start off as extremely large stars, and the bigger the star, the more likely it is to have at least one companion star. Remember how stars form – by clouds of gas and dust pulling together as a result of gravity. If a cloud is vast enough to produce a big star, it's more than likely able to form two or more. A very large star is about three times as likely to have at least one companion than a small star.

It's also true that the bigger a star is, the shorter its life will be, because the extra pressure from the gravitational pull of its greater mass means that it goes through the fusion processes faster. This means that a star that will end up as a black hole will have had a relatively short life. As a result, if its companion formed a little later than the black hole's parent star did, or isn't quite as big, you will end up with a black hole that is accompanied by another star, as is the case with Cygnus X-1. If, on the other hand, the stars become black holes at a similar time, we end up with a pair of black holes orbiting each other. This is also relatively common: we will visit one of these pairs later on in our trip and see how they made possible a breakthrough in Earth-based astronomy.

Interestingly, when John Michell first came up with his early version of the black hole concept in 1783, he accurately described how we could detect the existence of a black hole by its influence on light-emitting companions. He wrote: 'If any other luminous bodies should happen to revolve about the [dark stars] we might still perhaps from the motions of these revolving bodies infer the existence of the central ones with some degree of probability, as this might afford a clue to some of the apparent irregularities of the revolving bodies, which would not be easily explicable on any other hypothesis.'

We aren't going to visit an isolated black hole, as it's far less interesting to look at than an active pair like the Cygnus X-1 system. But even though we don't have the bright indicators of accelerated matter streaming inwards, or a distorted accretion disc, lone black holes are not impossible to detect.

Because of the strength of the gravitational field close to a black hole, it distorts the path of light coming towards it from behind (as far as our viewpoint is concerned). Just as light from the accretion disc of the black hole we've just seen is bent around it, so light from distant stars and galaxies will bend in towards the invisible star – the result is that it acts like a vast lens, which can produce a range of effects that will make it detectable.

Most dramatic is a so-called Einstein ring. This is when light from a bright source is originating exactly behind the black hole from the viewer's position. Light heading out from that source will be bent inwards symmetrically around the black hole, so what was a single point source of light will end up as a ring of light outside the black hole's dark shadow, where any light is bent back into the event horizon. You will also see stars a little off centre behind the black hole appearing more than once around the outside of the black hole's shadow as light travels around the black hole by different routes.

In principle, there is another way that isolated black holes can be spotted too, although in practice, even with today's instruments, this can only be done when a ship is already very near to the location of the black hole. This is a result of Hawking radiation.

Speculation alert

Although there is good reason to think that Hawking radiation exists, as of 2023, it had not been detected as it is so faint that it is unlikely to ever be identified from Earth.

Hawking radiation is a quantum effect, predicted in 1974 by the English physicist Stephen Hawking. According to quantum theory, empty space is seething with particles that fluctuate into and out of existence as energy levels are subject to the uncertainty principle, and over very short periods of time, the level in a location can climb high enough to spawn a pair of particles, with the sudden spike in energy very briefly converting to mass in good Einsteinian $E = mc^2$ fashion.

If this occurs, you get one matter particle and one antimatter particle, which in the normal course of events merge together and annihilate back to energy immediately. Antimatter might sound like science fiction, but it's a real thing. All matter particles, such as the electron, have an antimatter equivalent (in its case the positron). The antiparticle typically has the opposite value of some fundamental properties of matter, notably its electrical charge – and when a particle and its antiparticle come together, they annihilate as energy in the form of photons. In the case of these 'virtual particle' pairs that form in empty space, though, the extremely brief period of high energy will usually end so quickly that the particles are not normally detected.

What Stephen Hawking predicted was that when this happens very near to a black hole's event horizon, the particle pair could be divided. The particle left behind in the real world would have positive mass – but mass is something that is itself conserved, which means that, in effect, the particle absorbed by the black hole has to have negative mass, and the overall size of the black hole would shrink by an infinitesimal amount. According to this theory, over vast periods of time (depending on the size of a black hole), all black holes should eventually evaporate, even though nothing ever escapes from inside the black hole's event horizon.

Near to a Schwarzschild black hole, the Hawking radiation is detectable, if uninspiring by black hole standards. But here at Cygnus X-1, it is totally drowned out by the radiation pouring out as matter spirals into the black hole and there is no possibility of picking it up.

Where did the info go?

The prediction from the existence of Hawking radiation that black holes would disappear over time has led to a huge amount of head scratching among cosmologists. In the time frame we've been able to observe black holes up close, we don't have enough data to be sure that such evaporation really happens, because it's a very slow process. However, if we assume the gradual disappearance of black holes is true, there is a problem connected to the nature of information and whether the laws of physics allow us to ever irreversibly destroy it.

It's part of the very nature of black holes that what goes in never comes out. When Hawking radiation shrinks a black hole, it is not a matter of stuff coming back out of the object, but rather anti-stuff going in and destroying it. The radiation that is emitted just outside the black hole's event horizon is decoupled from the contents of the black hole. But as we will see in a moment, this gives a problem for quantum physics – the physics of the very small.

Usually, the gravitational action of big things like stars and the behaviour of tiny quantum particles can be treated entirely separately. Throughout the 20th and 21st century, scientists attempted to combine the effects of gravity and quantum physics, but the general theory of relativity is not a quantum theory – the two don't fit together. If they are to be united, something has to give.

The leading 20th-century theory that attempted to combine the two, string theory, has now been dismissed, while other attempts to provide a quantised theory of gravity, notably loop quantum gravity, have also fallen by the wayside, but nothing has yet successfully enabled a combination of gravitational and quantum theories.

Speculation alert

As of 2023, no attempts to produce either a theory adding gravity to the other, quantised forces, such as electromagnetism, or producing a specific quantised theory of gravitation has entirely been ruled out, but string theory and its successor M-theory were increasingly considered to be ascientific – impossible to either prove or disprove scientifically, and as such of little value in explaining the universe. Back then, it seemed possible that no combined theory would ever be produced, or that it would require a totally new way of thinking about the problem.

In most parts of time and space, there are very few problems arising from the lack of integration between theories of gravity and quantum physics because they deal with opposite extremes of the size scale, but in two aspects of cosmology – the big bang and black holes – the quantum world and the general theory of relativity bump up against each other extremely uncomfortably.

As we have seen, the nature of black holes – the very prediction of their existence – originates in the general theory of relativity. But Hawking radiation is a purely quantum effect. And the combination makes for an uncomfortable friction with one of the assumptions that lie at the heart of quantum theory. This hugely successful theory that lies behind all electronics, and much more, describes the behaviour of quantum particles, such as photons, electrons and atoms, using Schrödinger's equation. This gives us a way to predict how the state of a collection of quantum particles will evolve over time. Here, 'state' refers to the collection of properties of all the quantum particles involved, for example, the energy and position of each particle.

Famously, quantum physics is often regarded as weird because these properties like position are only knowable when an interaction of the particles with their environment occurs – before then, all that exists are a set of probabilities. We are not used to this with the position of, say, a tennis ball or a starship. This can make the whole quantum description of the universe feel fuzzy, because of the uncertainty involved. But it's important to be aware that the probabilities involved are *not* themselves fuzzy – they can be precisely calculated.

The outcome of all this is that there should be a set of information about an earlier state of the system of quantum particles that can be deduced from a later state. This information should be conserved, much as energy is usually conserved. However, the decoupling between the contents of the black hole and the energy loss as Hawking radiation means that somehow the information swallowed by the black hole disappears – and quantum theory tells us that this should not happen.

Note, by the way, that this ability to make deductions from available information is almost entirely a theoretical one. We can do it for very small systems of particles, but systems rapidly become far too complex, making the calculation impossible with anything that we would recognise as a normal physical object: this will be made up of vast numbers of quantum particles, many of them constantly interacting with each other.

Imagine, for example, that you have a book like *Interstellar Tours* in physical paper form. (We know that some of our passengers treasure these historical artefacts.) Let's say that you take your book and rip out one of the pages and burn it. After that, you rip out another page and eat it. In each case, *in theory*, if you had access to enough information on the final state of the quantum particles that originally made up those pages, and how they got to that state, you could recreate the original paper version, including all the information that they contained, both in the physical structure of the paper and in the words that were printed on them.

The real world is a very different prospect. Getting to the original information from the resultant pile of ash, or even worse, from the combination of excrement and atoms that may have been incorporated into your body from the page that you eat, would never be possible, however clever your technology was. So, this was never a practical issue. But such deductions of the information content are no longer even theoretically possible once a black hole gets involved in the transformation process.

The resultant confused state of understanding is known as the black hole information paradox. In the 21st century in particular, cosmologists produced a whole stream of ideas that supposedly solved the paradox. In some cases, this involved imposing a mathematical concept derived from string theory, known as the holographic principle, that effectively leaves the information floating at the event horizon of the black hole. It's worth unpacking this one a bit as it gets far more coverage than it really deserves.

The holographic view

We need to start with exactly what a hologram is. Imagine looking through a totally transparent window (or, indeed, one of the *Endurance*'s viewing walls) at a view of the world outside. Now, let's replace that window with an ultra-high-resolution photograph, so detailed that as far as your eyes are concerned it contains just as much detail. No matter how good the image is, it will be soon become obvious that it's not a real window.

The reason that you can tell the difference between the image and reality is that the photograph only reproduces photons of light that were coming in from outside in the exact direction that leads to the position of the camera. If you change your position in front of a window, the view that you can see changes. But if you move around in front of the photograph, everything in the image stays the same.

However, just imagine that your window was a special kind of photographic apparatus that could capture photons hitting the window from all different directions, then recreate those photons when the window photograph was viewed. Now, as you moved around, the view would change. You would have captured a true three-dimensional image on the flat plane of the glass. And that's what a hologram is. It uses a special recording technique involving interference patterns (more detail on these when we get onto gravitational waves) to capture information on all photons arriving at the glass, not just those heading towards a camera, then reproduces those photons.

When we move on to speculative astrophysics, the holographic principle treats the event horizon of a black hole as a special kind of holographic boundary, capable of holding (in some undetermined way) the information that would otherwise be lost in the black hole. Some get so excited about this principle that they will tell you that it is entirely possible that the whole universe is just a hologram. For example, American physicist Leonard Susskind once said, without a hint of uncertainty: 'The three-dimensional world of ordinary experience – the universe filled with galaxies, stars, planets, houses, boulders, and people – is a hologram, an image of reality coded on a distant two-dimensional surface.'

Unfortunately, there's a real problem with the holographic principle being applied to the universe as a whole. As German physicist Sabine Hossenfelder explained: 'The idea that the universe is a hologram only works if the cosmological constant is negative. The cosmological constant* in our universe is positive. And even if that wasn't so, the idea that the universe is a hologram would still not be supported by evidence. What does any of that have to do with real holograms? Nothing.'

Other attempts at solving the problem of black hole information loss worked by appreciating that a black hole might not actually be a true 'singularity' of zero size, and that perhaps the true nature of black holes would allow the information to eventually escape when the black hole entirely evaporates, something that we have never experienced.

It ought to be stressed that these ideas are not based on any observational data – they are all simply about playing with mathematics – in some cases, such as the 'Anti-de Sitter space' concept giving rise to the holographic principle, maths that doesn't even

---

* The cosmological constant is a component of Einstein's general theory of relativity equations that reflects the energy density of space. The concept of dark energy accelerating the expansion of the universe requires this constant to be positive.

apply to our universe's actual behaviour. This is the result of making assumptions to simplify the mathematics that we know aren't true in reality. But even when the maths is appropriate, the best we can say is that the numbers fit with what's observed.

Unfortunately, there is no way to distinguish between the different mathematical tricks all of which could fit with what's out there. This is not science, and never will be: it is recreational maths. We are still to make any observation that gets us any closer to what happens to the information that goes into a black hole – any ideas are still pure speculation and likely to remain so for a long time to come.

Going large

Cygnus X-1 is typical of what you might call a common or garden black hole – if a black hole could ever be considered ordinary. Such a black hole is the result of a large star collapsing and will have a mass of up to tens of times that of the Sun. But there is a whole different class of black holes that, by comparison, make Cygnus X-1 seem puny. We're on our way now to Sagittarius A*. This is a black hole at the centre of our galaxy, the Milky Way. And it is enormous.

Despite its name, Sagittarius A* is only just in the constellation of Sagittarius as seen from Earth – it sits on the border with the adjacent Scorpius constellation. And by supermassive black hole standards, this is an insignificant example – one of the smallest ever discovered. This apparent smallness may, however, be in part due to the fact that all the other known supermassive black holes are located in other galaxies, meaning that only the larger ones have been found because they have to be detected from a great distance. Due to the limitations of the hyperspace drive, we have never been able to visit another galaxy, so our own Milky Way's supermassive black hole is the only one we have been able to study in any detail.

To get an idea of what counts as small in supermassive terms, Sagittarius A* is about 4.3 *million* times the mass of the Sun. The biggest known supermassive black holes are around ten billon times the mass of the Sun. In terms of size, by now we're used to black holes being relatively small. But the monster in the centre of our galaxy is about seventeen times the width of the Sun. That's still tiny considering its mass – it would sit inside the orbit of Mercury, so it's smaller than many giant stars – but it is still impressively enormous for a black hole.

We will be dropping out of hyperspace soon. As a taster, we can see what the best view of Sagittarius A* looked like when it was first observed back in the 2020s.

View 24

Sagittarius A*

To see outside the ship, use the QR code or
visit www.interstellartours.co.uk/view24.html

This is an image based on radio waves, converted into the optical range to make it visible – Sagittarius A* is not visible from the Earth as there is so much dust between our planet and the galactic centre that the light from its accretion disc can't make it through.

It would be nice to think that all supermassive black holes were identified using a simple naming convention, with the '*' indicating their nature – but as is often the case in astronomical naming, simplicity doesn't come into it. The 'A' indicates that this is the brightest radio source in the constellation (the black hole was first discovered as an unknown radio source) and the '*' simply indicates that it's an interesting feature. Some other supermassive black holes have also been given the '*' designation, but often they are simply referred to by the name of the galaxy that surrounds them.

One of the ways that information was first accumulated about our home galaxy's supermassive black hole was by observing a collection of stars that orbit it, just as planets travel around a solar system. These stars are unimaginatively named S1, S2 and so on. Even now, we don't have a total for how many there are as new ones keep being spotted, but there are well over 60 of them. We should now be able to see some of those orbiting stars in a repeating accelerated image of their motion (each cycle is around twenty years' worth of movement).

View 25

Sagittarius A*
S stars

To see outside the ship, use the QR code or
visit www.interstellartours.co.uk/view25.html

The S stars are remarkable both for their location at the centre of the galaxy and their speed of orbiting. S2, for example, which was the second-closest known when these stars were first classified, can get as close as 120 AU to the black hole – that's just 40 times the distance Neptune is from the Sun, which is remarkably close for a star to orbit something as hefty as a supermassive black hole. As a result, its orbital speed can be as high as 7,700 kilometres (4,800 miles) per second – which is 2.5 per cent of the speed of light.

Galaxy central

The Milky Way is a well-established galaxy (we'll take a big-picture look at the Milky Way later in the tour), and our very own supermassive black hole has long since cleared any nearby material, so it is relatively dark. Of course, when we're dealing with something on this scale, 'relatively dark' can be misleading – it is still hundreds of times brighter than the Sun. But the theoretical maximum limit for the brightness of such a black hole is around 26 trillion times the Sun's brightness. Random material passing by does sometimes head in the wrong direction and make it into Sagittarius A*, but most objects, like those orbiting stars, are no more at risk of destruction than the Earth is of plunging into the Sun – which is a risk that is very close to zero. As with Cygnus X-1 you will see a classic black hole's distortion of the light output of the accretion disc due to the angle we're viewing it.

View 26

Sagittarius A*

To see outside the ship, use the QR code or visit www.interstellartours.co.uk/view26.html

However active they are, it does appear that almost every galaxy has a supermassive black hole at its centre. The two go together in what could be considered a symbiotic relationship if they were alive. And some of those other supermassive black holes are far more energetic than our own, because their galaxies are far younger than the Milky Way. We can see plenty of young galaxies, as telescopes double as time machines for seeing into the past. Because light takes time to reach us, the further away something is, the further back in time we see it. We see our nearest major galactic neighbour the Andromeda galaxy, for example, at a range of 2.5 million light years as it looks as it was 2.5 million years ago. With modern telescopes we can see back billions of years.

For a long time, astronomers were puzzled by objects in the sky that were both very bright and very far away. They looked like stars – points of light – but clearly weren't stars as they were much further away than any individual star could be seen. As a result, they were given the name 'quasi-stellar objects' – better known as 'quasars' for short. With the discovery that supermassive black holes were a common feature at the heart of galaxies, it was realised that quasars are the supermassive black holes of young galaxies, which are still consuming enormous amounts of matter*, and as a result, producing vast amounts of high-energy light as the matter is accelerated towards the black hole.

---

* Whenever we deal with time travel, even the virtual time travel viewing offered by telescopes, verb tenses can get a little mangled. When we say that quasars 'are' the supermassive black holes of young galaxies that 'are' still consuming vast amounts of matter, what we really mean is they *were* in this state billions of years ago, as we see them when we *are* viewing them today. There really should be a special time travel tense, enabling us to say that they 'ware' like this.

We can't take you to see a quasar, as they are beyond the range of our ship (and the majority of quasars are beyond us in the distant past too), but we can take a look at an artist's impression of the clumsily named ULAS J1120+0641. The ULAS part tells us it was discovered by the UK Infrared Deep Sky Survey (UKDISS) and specifically by the UKDISS' Large Area Survey programme (ULAS), with the initials followed by the coordinates in the Earth's sky, placing the quasar visually in the constellation of Leo. This is a distant galaxy, around 28.9 billion light years from Earth, seen as it was around 13 billion years ago.

View 27

ULAS J1120+0641

To see outside the ship, use the QR code or visit www.interstellartours.co.uk/view27.html

If may look like there was an error in the number above, positioning the quasar 28.9 billion light years away, but the expansion of the universe makes this kind of oddity commonplace when dealing with great distances in space. You might expect light to take 28.9 billion years to cover 28.9 billion light years. But during the 13 billion years since this light set off, the universe has become much bigger. As a result, when we look at very distant objects, they are much more distant than the timescale suggests.

The visible universe size – the distance we could in principle see if there were no limits on the ability of light to get to us – is around 45 billion light years in each direction, making the visible universe about 90 billion light years across. In principle, the universe could be far bigger, or even infinite, but we have no way of ever seeing beyond the (still expanding) limit of the size of the visible universe.

Back to that quasar ULAS J1120+0641, it's hard to get your head around the sheer power output of this object – it's about 63 *trillion* times as bright as the Sun, brighter than a whole galaxy's worth of stars, and it is estimated that the black hole responsible for it is around 2 billion times the mass of the Sun. In the artist's impression, you can see the familiar intense beam of a jet, funnelled by the magnetic field produced as the spinning black hole pulls space and matter around with it. The jet consists of matter accelerated to near light speed, generating very high-energy light up to the X-ray and gamma ray region – in this case, far more dramatic in power and scale than anything we've seen in our galaxy. If anything came within light years of the quasar along the path of that jet it would have been fried.

Where did the supermassive black holes come from?

Most large galaxies are known to have supermassive black holes at their centre (and even where they aren't known, they may well be there), suggesting that there is a direct connection between the two. What's more, in many galaxies there is even a relationship between the size of the central black hole and the amount of matter that's packed into the central region of the galaxy. Inevitably, when it was realised that supermassive black holes and galaxies usually co-existed, there was a chicken and egg debate of which came first.

All the evidence is that rather than an existing supermassive black hole causing the galaxy to form, or an existing galaxy growing itself a central black hole as a kind of celestial beauty spot, the two grew together, each influencing the development of the other. Material plunging into an expanding black hole can act as a stabilising energetic influence of the early structure of a galaxy, while the growing collection of material that will eventually form both the galaxy's stars and its unstructured gas and dust would help feed that central monster.

We think the first seeds of early galaxies can be seen in the remarkable imagery of the cosmic microwave background radiation. As we have seen, the universe has been expanding since the big bang. In the very early days, when matter first formed from the initial energy present, it would have been in the form of electrically charged ions, which prevent light passing through. But when the universe was about 380,000 years old, uncharged atoms formed and the universe became transparent.

At that point, ultra-high-energy gamma rays began to cross the universe. As the universe has expanded, the wavelength of that light has reduced due to the photons reducing in energy, to the extent that it is now in the form of microwaves. On Earth, these come from every direction in the sky – a faint background hiss known as the cosmic microwave background radiation or CMB for short.

Since the 1960s, when US physicists Robert Wilson and Arno Penzias first discovered the CMB, it has been mapped out using more and more sophisticated satellites. The result is the kind of omnidirectional plot of radiation that you can see on the viewing wall.

View 28

CMB plot

To see outside the ship, use the QR code or
visit www.interstellartours.co.uk/view28.html

This squashed oval represents the whole sky in every direction
from the Earth but projected onto a flat surface, an inverted version of the way that a map of the world requires such a projection.
The speckling shows variations in the intensity of the radiation.
The fact that we were able to detect this way back in the 20th
century is remarkable, as the variation in intensity between the
brightest and darkest points of the CMB is just 1 in 100,000.
What it enables us to see is a variation in the distribution of
matter – almost all hydrogen atoms – in the early universe. This
has been described as the 'echo of the big bang', but it would be
more realistic to call it the foetal scan of a baby universe.

The brighter parts show locations where the earliest galaxies
might eventually have formed. Over time, as matter was accumulated by gravity, we would have seen the formation of stars
and the coming together of galaxies or galactic clusters, with
attendant supermassive black holes growing alongside them.
These may have formed from the death of the earliest stars in
the proto-galaxy, or directly from the collapse of a large amount
of gas in the early years of galaxy formation.

It might seem that a supermassive black hole is the ultimate that space can offer us in the way of spectacle. But the

tour of the *Endurance* never fails to give us new possibilities for awe-inspiring vistas. Our penultimate voyage will be to some of the most catastrophic events in the universe – when black holes, and other massive bodies collide, sending vibrations outwards that shake the very fabric of space and time.

COLLISIONS 8

Black hole buddies

We have already visited a system with a pair of stars, one of which is a black hole – Cygnus X-1. Now we are arriving at another binary system where both of the stars have become black holes.

For a few million years, the two original stars would have orbited each other in a stable fashion. Remember, binary stars are very common in the universe and usually get along just fine. But then a sequence of events occurred that destabilised the system. First, the larger of the two stars, which typically means that it had a shorter lifespan, went supernova, blasting off a considerable amount of its mass.

We tend to think of orbits in terms of a smaller body travelling around a (relatively) big body. The Earth around the Sun; or the Moon (or an artificial satellite) around the Earth. But in reality, what happens is that both bodies are in orbit around their mutual centre of mass. This is a point between the centres of the two bodies where you could imagine the two bodies balancing on a seesaw. The centre of mass will be located closer to the more massive body.

In a system like that including the Earth and the Sun, that centre of mass is inside the Sun, because the Sun is much more massive than the Earth – although it's not quite the centre of the Sun itself, so the Sun does wobble a little as a result of the orbiting Earth. Even so, it looks like the Earth is simply in an orbit around the Sun*. When the system consists of a pair of stars of similar mass, they will both orbit a centre of mass that will be located roughly at the midpoint between them.

When the first star goes supernova, its mass drops considerably, and so the centre of mass moves towards the other star. During the period of change, the orbits will take a considerable amount of time to settle down, as matter blasts out from the supernova. When the orbits do stabilise, the star and the black hole will now be significantly closer to each other. Then, the second, smaller star will have gone supernova. As the two bodies were now already relatively close, the material blasting out in the explosion would disrupt the orbits considerably more, while the two black holes get closer still to each other.

In principle the pair could settle down to stability, but in practice, it would usually be the case that a system like this never recovers from this event. There is sufficient disruption that the two black holes end up spiralling in towards each other, becoming more and more unstable in the process. The eventual outcome is a dramatic collision. Black holes may be small, but bear in mind that each will still be more massive than the Sun.

That's what is about to happen where we have just arrived, although we are a little early for the collision. Looking out on

---

* The reality is more complex, of course, because the Earth is not the Sun's only planet and the others, particularly Jupiter, make their own gravitational contribution. For that matter, the Moon is relatively large compared to the Earth and also makes a difference.

the viewing wall there would be nothing to see except the distortion of light from other stars as the rotating black holes eclipse them, and their gravitational pull warps spacetime to change the light flowing past them. At this stage in their lives, there is no other material left in the system to flow into them. The black holes are truly dark.

Even so, the black holes are producing a series of waves on top of the Hawking radiation. Near to each black hole, spacetime is being warped to a high degree. As the black holes spin around each other, this warping is continually changing, sending out cyclical vibrations through the very fabric of spacetime. These waves travel outwards at the speed of light and can pass through anything and everything.

Back in 1916, shortly after concluding his work on the general theory of relativity, Einstein realised that because spacetime itself is a medium that can be stretched and twisted by the effects of gravity, it should be possible to send waves through it, rather like sending ripples through something stretchy like a Slinky spring. However, although he thought such waves would exist, he also decided that they would be impossible to detect as they would be extremely faint, requiring both extremely sensitive detectors and a massive distortion in spacetime to send a detectable wave through the universe.

The strength of such gravitational waves is dependent on three factors: how massive the moving bodies producing the waves are, the strength of their interaction and how far away they are from the detector. Everything that moves with a repetitive motion – such as the Moon orbiting the Earth – will produce gravitational waves. But to be picked up from a remote detector, we need bodies with stellar masses, ideally moving very quickly. This makes our pair of black holes, spiralling into each other faster and faster, an ideal source.

If you think of the pair of black holes a bit like a food mixer, stirring a thick mixture, as they spiral in, getting faster and faster, they send out ripples through the 'mixture' of spacetime itself that will increase in intensity and frequency, producing a distinctive pattern that gravitational astronomers refer to as a 'chirp', which is then followed by a rapidly fading 'ringdown' as the black holes merge into each other and shudder their way into a single body.

Back on Earth, despite Einstein's doubts, scientists started the search for gravitational waves during the 1960s, but it was not until 2015 that the first near-certain gravitational wave detection was made.

Spotting the invisible

The whole business of detecting vibrations in the fabric of spacetime is fraught with difficulty. Gravitational waves do not produce any effect that we can see directly. In fact, there are three issues that the gravitational wave astronomer must face: finding a way to detect such tiny shifts in the structure of reality, working out the direction that the wave has come from and distinguishing a gravitational wave from another kind of vibration that could trigger the detector.

Of these problems, the simplest aspect is working out the direction of the source, although it still requires some extra thought. A conventional telescope can only 'see' in one direction, even if it isn't using visible light. But waves in spacetime aren't blocked by physical objects. So, although, as we'll see, a gravitational wave detector can have some directional capabilities, it can't distinguish, for example, a wave that comes from above and one that comes from below, passing straight through whatever the detector is mounted on (such as the Earth).

Astronomers have two ways to get around this. They can use conventional telescopes pointing in the possible directions that the gravitational waves could have originated from, attempting to detect some new electromagnetic disturbance caused by the same phenomenon as produced the gravitational waves. This isn't definitive proof but does provide a good indication. Alternatively, they can use more than one gravitational wave observatory, each sited at different physical locations, using the difference in time of arrival of the signals to work out the direction that they came from.

Then there's the matter of false alarms. This is nothing new. Users of traditional telescopes have sometimes mistaken other sources of light for astronomical bodies. And this problem became considerably worse when radio telescopes were introduced. Instead of producing anything like a picture, the early radio telescopes simply recorded a series of numbers, corresponding to the frequency, strength and direction of the radio signal. But such data could also be generated by more local radio sources.

Radio telescopes located near domestic housing have been known to have to provide suppressors – clip-on devices that reduce radio noise – to people whose vacuum cleaners or other household devices were producing unintentional radio interference. It was only by recognising the structure of typical terrestrial signals that radio astronomers could necessarily eliminate misleading data. In the case of gravitational waves, because of the way they are detected (more on that in a moment), they are particularly susceptible to picking up stray vibrations – but not the good variety, beloved of the Beach Boys.

Bad vibrations

In a moment we'll look at how successful gravitational waves detectors work (and why the early attempts failed). But all such detectors depend on the fact that ripples in spacetime should have an impact on physical objects, particularly if those objects have plenty of concentrated mass and are carefully suspended so they can move as a result of the waves passing through. The false alarm problem is that the detectors can also be set in motion by local conventional vibrations.

One obvious possible cause for concern is earthquakes. Even in very stable regions, small earth tremors are relatively common. And a big earthquake will produce vibrations detectable by seismometers anywhere on the planet. But a gravitational wave detector is, in effect, an incredibly sensitive seismometer. It's not intending to look for movements in the Earth, but in spacetime itself – but that won't stop the shaking of the ground having an influence on the detector that would be difficult to distinguish from a genuine gravitational wave in isolation.

Gravitational wave detectors must be so sensitive that even a passing car a few miles away will generate sufficient vibrations to produce a signal. Apart from placing detectors in isolated locations, there are two approaches that have been used together to minimise the chance of a false signal. One is to incorporate seismometer data into the information collected by a gravitational wave observatory, to be able to eliminate the more obvious ground-based vibrations. The other is to have at least two detectors, situated thousands of miles apart. Although it's just possible that a remote earthquake would be positioned so it's a similar distance (and hence timing and strength) from both detectors, more often than not, this wouldn't happen, and the difference in the signal at the two gravitational wave detectors will pick it out as a false positive. With even more detectors in play, most of the local (in the astronomical sense) vibrations can be eliminated.

The ultimate challenge

The first attempt to pick up gravitational waves dates back to the 1960s, when American physicist Joseph Weber set up an experiment using a solid two-tonne cylinder of aluminium, around two metres (6.6 feet) long and one metre (3.3 feet) in diameter. It was suspended from shock-absorbing supports to minimise local vibrations and had a number of quartz crystals attached to it. These weren't for decoration, but to measure any expansion and contraction of the cylinder.

Quartz (crystalline silicon dioxide) is a piezoelectric material, which means that it generates an electrical current when it is squeezed or stretched. Weber hoped that the gravitational waves would result in the bar being stretched and squashed and this would produce a current from his attached crystal detectors. Unfortunately, the equipment was far too insensitive to pick up anything other than local vibrations – at the time, Einstein's prediction of failure proved to be correct.

Weber increased the sensitivity of his devices (although we now know they were incapable of ever detecting a gravitational wave) and also set up a second device, with the two cylinders separated by around 1,000 kilometres (600 miles). This was intended to help eliminate the local vibration problem. And in 1969, it looked like Weber's work had paid off. He picked up a near-simultaneous signal at the two detectors. But even this wasn't enough to say that he had discovered gravitational waves.

There was another problem to overcome that would require some mathematical manipulation. And as a result, the search for gravitational waves moved from the traditional astronomical approach of directly observing some kind of image or signal to a methodology more common in fundamental physics, where all that's ever possible to say is that there is a certain level of statistical confidence that the searched-for event had happened.

Statistical astronomy

Weber might have received two signals at around the same time, but coincidences happen all the time (despite the much-repeated mantra of TV detectives: 'I don't believe in coincidences'). If we imagine that there were no gravitational waves, or that Einstein was right and it was impossible to detect them, there would still be a range of reasons, from local vibrations to errors in the electronics, that could lead to a detector picking up an apparent signal. If each detector had these occasional false positives, randomly distributed over time, sometimes, over a long enough period of observation, at least one of the random observations at one detector would coincide in time with a random observation at the other. Some simultaneous detections would indeed be pure coincidence.

Weber came up with an approach that would make it possible to estimate the chance of a simultaneous detection happening due to such a random coincidence, rather than as a result of the two signals sharing a cause. (Strictly speaking, as we have seen, this shared cause could be a seismological event that was a similar distance from both detectors, but seismographs could eliminate that possibility.) To see how Weber did this – an approach that would continue to be used in a more sophisticated form in all future gravitational wave detection – imagine you had two rolls of paper on which measurements of movements in a pair of detectors were recorded as blips in a continuous line.

What Weber did (mathematically rather than physically) was to slide one of the long pieces of paper along the other, noting when pairs of blips on the two charts lined up. When this happened, there definitely wasn't a shared cause, because these vibrations had not occurred at the same time. But Weber was able to calculate the chances of a coincidental detection occurring with no cause by the frequency with which his unconnected signals lined up with each other. This process was given the name of a 'time slide'.

The result of Weber's analysis was to suggest there was indeed a shared cause for his 1969 joint detection – it was unlikely there would be such a coincidence. Unfortunately, though, even unlikely events can happen. The chances are tiny that anyone would win a major lottery, but it happens on a regular basis. And Weber's analysis was not up to the kind of statistical standards that would later be required. More importantly, Weber's apparent success triggered the construction of a whole range of similar detectors at different sites around the world to attempt to duplicate it.

A real advantage over later gravitational wave detectors was that Weber's bars were simple pieces of kit, easy enough for a university to put together. A detector was even taken to the Moon in 1972 by the Apollo 17 mission, as an ideal location for getting away from distracting unconnected vibrations. (Admittedly, the Moon-based device was even less sensitive than Weber's, as the Apollo lander did not have the capacity to carry a two-tonne bar.)

No one else obtained the same results. In scientific terminology, the experiment was not reproducible. Although in principle Weber could have succeeded, the more that others failed to detect gravitational waves, the less likely was his 1969 success. We now know that his detectors were definitely not sensitive enough to produce a real result. Soon after Weber's results, theoreticians demonstrated how problematic this was. English astrophysicist Martin Rees (later Astronomer Royal) calculated that to be detectable by Weber's devices, the waves would have to be so strong that they would have ripped the Milky Way galaxy apart – so it's just as well that his detection was an error.

A successful detection would require a totally different – and far bigger – piece of technology.

Monstrous machines

Strangely, there would be confirmatory evidence for the existence of gravitational waves well before they had ever been
detected. This was because of measurements on a weirdly behav-
ing pulsar. As we discovered when we visited one, a pulsar is a
rapidly spinning neutron star, with a clock-like consistency in
the frequency of its radio pulses. But in 1974, American radio
astronomers Russell Hulse and Joseph Taylor found a pulsar that
seemed to be speeding up and slowing down every few hours.

The astronomers deduced that the pulsar was orbiting
another star (probably another neutron star), which would
account for the variation in the apparent pulse rate, as there
would be a Doppler effect shift in its radio frequency, depending
on whether the pulsar was moving towards or away from the
Earth. In itself, this said nothing about gravitational waves. But
it was also the case that such a system would be expected to
produce gravitational waves. And that takes energy. The result
of this leaking of energy should have been that the two stars
gradually got a little closer to each other.

As the stars orbital distance got lower, they would have to
speed up in their orbits to avoid crashing into each other. And
that would mean that the pulsar would change direction more
frequently over time as seen from Earth. To observers, the frequency with which the pulsing was varying should also have
gone up. And that is what was observed. The result wasn't
definitive. In principle, something else could have caused the
decaying orbits. But what was observed was consistent with the
existence of gravitational waves – and that was enough to win
Hulse and Taylor the Nobel Prize in 1993.

However, this feat was eclipsed a couple of decades later
on 14 September 2015 by the first true detection of gravitational waves. Doing this involved a massive pair of

detectors, collectively known as the Laser Interferometer Gravitational-Wave Observatory (LIGO). The two sizeable structures were located in America at Livingston, Lousiana and Hanford, Washington – over 3,000 kilometres (1,865 miles) apart.

We can see historical images of the LIGO observatories on the viewing wall now.

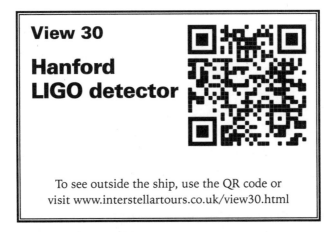

As you can see, each detector had a pair of long tubes running from the central site at right angles to each other. These metal tubes were four kilometres (2.5 miles) long and 1.2 metres (3.9 feet) across, with practically all the air pumped out of them. A pair of laser beams were sent down each tube, making repeated journeys as the beams reflected off mirrors at each end before being brought together to make an interference pattern.

Interference is an effect that you see whenever similar waves overlap. Imagine dropping two stones into a still pool of water and watching the waves from the stones expanding outwards until they meet each other. At some points, the surface of the water where the sets of waves from both sources overlap would be heading upwards at the same time. These waves would reinforce each other, making the overall wave at that point bigger. The same thing would happen if both waves were going down at the same time. But if the water going down as a result of one wave met water going up from the other wave, the pair would cancel each other out. This process, producing a pattern of stronger and weaker vibrations, is known as interference.

If the waves continue to be produced, there will be a set of locations with strong up and down movement occurring and a second set of locations with hardly any movement. But if one of the wave sources moves a little bit, these locations will change. The interference pattern shifts – and a very small shift in the initial sources can produce a much bigger shift in interference pattern.

Similarly, if one of the arms of an interferometer detector like LIGO is extended or shrunk by gravitational waves, the interference pattern produced when the two laser beams meet will shift. And because those beams pass up and down the arms multiple times, an extremely tiny change in the distance travelled will result in a noticeable shift in the pattern.

In detectors like LIGO, there are very heavy mirrors at the end of each long arm, suspended to avoid vibration, so that those tiny changes from the gravitational waves will shift the pattern, while the impact of local vibrations is minimised.

You can see one of the LIGO mirrors here. The suspension system had four pendulums to damp out vibration, holding the 40-kilogram (88-pound) mirrors on glass fibre supports the thickness of two human hairs.

The real thing

What LIGO detected on 14 September 2015 is what we are about to witness outside the ship. We'll first see the final seconds of a pair of black holes about to merge in enhanced visible light.

While you are watching this remarkable event, our gravitational wave detectors are mapping out and recording the influence of the black holes on spacetime, showing distortions in spacetime as orange ripples, resulting in gravitational waves shown in purple.

View 33

Black hole merger: gravitational waves

To see outside the ship, use the QR code or visit www.interstellartours.co.uk/view33.html

Unlike our predecessors, we can witness the production of gravitational waves directly at their source. But like Joseph Weber with his bars, the LIGO scientists had to rely on statistical analysis, using time slides, to establish the chances that the observed results were due to a single, non-local source, assumed to be gravitational waves, rather than a local source, or the result of coincidence. The measure used was designed to only make a detection a serious contender if it would happen by chance once in 10,000 years or longer.

To make matters even more nail-biting for the scientists involved, when apparent gravitational wave events were recorded by LIGO, they could have been intentionally faked signals. To see if the system was working, the design included the possibility of producing 'blind injections' – fake data that would be pumped into the system that simulated how a real detection was expected to look. Whether or not such blind injections took place was kept secret from those making the observations. It would only be after they decided they had recorded a real gravitational wave event that they would check to see if a blind injection had been made.

The point of a blind injection is that the validity of any detection can be influenced by decisions made by those operating the experiment. Humans must decide whether to ignore signals that may have been caused by a local event. For example, if a plane was known to have flown low over one of the detectors, or there was an earthquake, this part of the data would be flagged up as unreliable. But the decision to ignore data could result in it being mishandled and a genuine signal being ignored. So the blind injections, which should be identified as potential real detections, were a useful test of the effectiveness of the human part of the process.

This wasn't just about local vibration issues. The mechanism of using time slides also has some human input into how the measurements are taken. It is down to the scientists to decide the size of the increments by which they move one set of data against the other to look for coincidences. Similarly, someone must identify how long a sequence of data will be compared. Each of these choices could change the interpretation of the data, giving a potential for missing genuine detections that blind injections would highlight.

This was not just a hypothetical occurrence. In the early testing of LIGO, what appeared to be a detection occurred in 2007. This caused a lot of excitement, shifting everyone's focus to analysing the data and making sure it was robust. It was only then that the signal was revealed to be a blind injection. But so strong had been the focus on this first event that a second blind injection was entirely missed.

There was no such problem with the 2015 detection.

A remarkable feat

Given how far technology has advanced since, it's hard to imagine how the scientists of the 21st century would have been capable of detecting gravitational waves – it was a truly remarkable achievement. Kip Thorne from the California Institute of Technology, one of the Nobel Prize winners for the detection, put together a hierarchy of scale to demonstrate just how small the vibrations that had to be detected were.

His starting point was a centimetre (around 0.4 inches) – 1/100th of a metre and an easily visualisable size, about the width of an adult's finger:

- Divide that by 100 (1/10,000th of a metre) and you get the thickness of a human hair;
- Divide that by 100 (1/1000,000th of a metre) and you get the wavelength of visible light;
- Divide that by 10,000 (1/10,000,000,000th of a metre) and you get the diameter of an atom;
- Divide that by 100,000 (1/1000,000,000,000,000th of a metre) and you get the diameter of the nucleus of an atom;
- Divide that by 100 (1/100,000,000,000,000,000th of a metre) and you get the largest motion expected between the mirrors of a gravitational wave detector like those used in LIGO. For those unfamiliar with metres, that's 1/3,937,000,000,000,000,000th of an inch.

The LIGO detectors had to pick up a movement in the mirrors that was 100 times smaller than the nucleus of an atom. And they succeeded. Of course, the LIGO scientists didn't have the option of witnessing the death spiral of a pair of black holes directly on the starship *Endurance*. But they were able to show a pair of signals from the two detectors that behaved exactly as the expectation would be for a pair of black holes with masses of around 29 and 36 times the mass of the Sun spiralling into each other and colliding.

We can take a look at the cleaned-up signals from Livingston and Hanford, placed alongside each other.

View 34

LIGO first detection

To see outside the ship, use the QR code or visit www.interstellartours.co.uk/view34.html

What we see here is the last moments of the black holes spiralling into each other, after the peak comes a 'ring down' as the black holes combine and vibrate together before settling down to a stable, single black hole. Before reaching the Earth, those gravitational waves had crossed around 1.4 billion light years of space, arriving on the outskirts of the Milky Way around 50,000 years before they were detected.

After the initial LIGO observation, gravitational wave astronomy would go from strength to strength, with the addition of extra detectors around the world adding to both the ability to make detections and to pinpoint the origin of gravitational waves. This capability was enhanced much further by the addition of LISA and its successors to the gravitational wave astronomers' toolbox.

Launched in 2037, LISA (Lisa Interferometer Space Antenna) had arms that were not four kilometres long, but a million kilometres (around 620,000 miles). This would not be physically possible on Earth, but LISA consisted of three satellites, positioned in space to provide the mirrors for a much bigger equivalent of LIGO. The 'arms' here were the three gaps between each satellite.

Speculation alert

The LISA project has had something of a chequered past. The joint NASA/European Space Agency (ESA) experiment was originally intended to have gone live by 2010, but it was delayed when NASA pulled out of the experiment, leaving the ESA to manage it alone. The expected launch date of the satellites was then put back several times, from the late 2020s, to 2034. By 2023, it had slipped again to 2037. It would not have been surprising if it had been put back again or even cancelled.

The first gravitational wave detection was of an event far outside our own galaxy. And it's only fitting that for our furthest voyage of the whole tour, we now get under way to travel outside of the Milky Way itself, to see our galaxy as a whole. We can't reach our nearest full-sized galactic neighbour, the Andromeda galaxy, which is about 2.5 million light years away because that's beyond our ability to manipulate hyperspace, but we can get out of the galactic plane and take a look at our wider home.

MILKY WAY 9

Welcome to our island universe

When galaxies were still speculative as a concept, they were sometimes referred to as island universes – a beautifully poetic description. It's certainly one that holds up in establishing the sense of isolation we have between our galaxy, the Milky Way, and the rest of the universe. The observable universe is estimated to contain in the region of 200 billion galaxies (this has to be a distinctly back-of-an-envelope estimate – there is no practical way to count them all) – so the Milky Way represents only a tiny fraction of the whole universe – yet from the viewpoint of human beings, it's an impressively big place. Our galaxy is home to over 100 billion stars and their associated planets, not to mention 5×10^{37} tonnes of gas* floating in between those bright points of light.

Just as is the case with stars, galaxies vary hugely – as do different regions of the universe. In some places, the galaxies

---

* Statistics like 5×10^{37} tonnes of gas can be hard to envisage. This is roughly the equivalent of the mass of 25 million stars the size of the Sun.

are relatively crowded together. Elsewhere there are gaps as much as 100 million light years across – more than a thousand times the diameter of the Milky Way. Although cosmologists tend to pretend that the universe is pretty much the same in every direction, it is actually surprisingly heterogenous. Some galaxies are much smaller than the Milky Way with perhaps only around 100 million stars inside – others are bigger, although the Milky Way is no minnow, sitting up in the top few per cent of galaxies for size.

When we look at a galaxy like the Milky Way – which we will do in a moment – it's easy to think of it being crammed tightly with those many billions of stars. From a distance, it looks as if there is hardly any empty space left inside, given all that stellar matter. But the reality of a spiral structure that is around 90,000 light years across is that there is plenty of room to spare. Although making use of hyperspace means we don't encounter obstacles as we travel, if we were to blast through the Milky Way at near light speed it's only occasionally that we would be at risk of colliding with a star. Typically, stars in the Milky Way are trillions of kilometres apart*.

We are now positioned as far as can be achieved with the hyperspace drive – far enough out from the Milky Way to see it as a whole entity, yet still well isolated from our nearest significant galactic neighbour, the Andromeda galaxy, a similar-sized galaxy to the Milky Way. That distance is gradually reducing – the two galaxies are on an inevitable collision course – but they

---

* The space between stars in the Milky Way doesn't make the earlier observations about the dangers of high-speed travel without using hyperspace incorrect. Although the risk would be relatively low of hitting a star, there is enough gas and dust out there that a near-light speed ship would almost certainly have a calamitous collision with something.

will not merge for another 4 billion years, so it's not a great concern for now.

Looking out on the splendour of the Milky Way, we obviously can't see Earth and the Sun among those 100 billion stars in that roughly 90,000 light year-wide structure. The Milky Way has a distinctive shape with two major spiral arms and a number of minor ones around a central bulge and accompanying bar shape, with its stars mostly in a relatively shallow plane around 1,000 light years thick (apart from the bulge), making it disc-like in form. Such spiral galaxies make up about 60 per cent of the galactic population, while the remainder, known as elliptical galaxies, are more like balls of stars, either spherical or ovoid in form.

Speculation alert

It is only very recently that we have been able to see what the Milky Way looks like from the outside. The view from Earth only shows a curved band of light. Prior to having an outside view, astronomers made projections of the shape of the Milky Way based on other galaxies they could see, like the Andromeda galaxy, and what they could deduce from the Earth-bound view of the Milky Way. Back in the 2020s, the detailed structure (for example, the number of spiral arms) was often stated as if it were fact. However, there was still considerable debate among astronomers in the 2020s as to the detailed structure of the galaxy, particularly in the spiral arms.

Spiral galaxies like ours typically contain more of the younger, brighter stars – often at the white/blue end of the spectrum – than ellipticals, and are also home to far more of the dramatic nebulae that act as stellar nurseries. By contrast, the ellipticals, which are often clustered together, are mostly made up of older, redder stars and include far fewer nebulae.

View 36

Milky Way showing Earth location

To see outside the ship, use the QR code or visit www.interstellartours.co.uk/view36.html

Despite not being able to pick out the Sun, we can point out our approximate home location in one of the spiral arms, known as the Orion arm (it seems Orion will remain the most significant of the constellations that act as companions for this voyage). If you were the sort of person who delighted as a child at providing a full address that ended 'the solar system, the Milky Way, the universe', you might enjoy adding a couple of neighbourhood features, making that 'the solar system, the Orion arm, the Local Fluff*, the Local Bubble, the Milky Way ...'.

Along with most of the other stars in the Milky Way, our home is swirling around somewhere we've already visited, the supermassive black hole Sagittarius A*, at a dramatic 725,000 kilometres (450,000 miles) per hour. Despite that speed, it takes the Sun around 210 million years to make a complete circuit around the galactic plane.

A wondrous band of light

Although any idea of looking at the Milky Way from the outside was a matter of science fiction until recently, it doesn't mean that the Milky Way was an unknown concept. It had been observed throughout history. We are used to science progressing with the centuries, but, oddly, the ability of most people to see the Milky Way is one that has become harder as time has passed.

To anyone from the pre-industrial age, the Milky Way would have been an inevitably familiar part of the night sky. But as towns and cities adopted streetlights, the relatively faint Milky Way faded from view across wide areas, leaving it only clearly visible from locations where the skies are still truly dark. We can see what the Milky Way looks like from Earth in the image on the viewing wall, but bear in mind that all such images tend to be enhanced compared with the reality of what the eye can see.

---

* More formally known as the Local Interstellar Cloud, the Local Fluff is the affectionate name of the locale of our solar system, a volume of about 30 light years across.

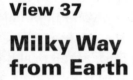

View 37

Milky Way from Earth

To see outside the ship, use the QR code or visit www.interstellartours.co.uk/view37.html

Cameras can always enhance the night sky by using a long exposure, enabling more photons to be collected than is possible for the human eye. The Milky Way as seen from Earth can be impressive, but even in the darkest skies, it does not look quite as dramatic as we see here. It wasn't until Galileo viewed the Milky Way through a telescope in 1610 that it was clear that it was a band of stars, rather than just a bright streak in the sky.

We get the unusually poetic name of our home galaxy from Ancient Greek mythology. It was said that this light-coloured band in the sky was the breast milk of Hera, Zeus' wife and the goddess of families. Zeus is said to have attempted to get his half-human son Heracles (who become the more familiar Hercules in Roman mythology) to drink Hera's milk to ensure that he became immortal. Zeus' attempt to do this while Hera was sleeping failed and she detached the baby, spilling some milk. 'Milky Way' came to us via Latin (where it was the *Via Lactea*), which is derived from the Greek *galaxias kyklos* (actually more accurately translated as a milky circle) – which is also where the word 'galaxy' originates.

Incidentally, apart from this streak in the night sky and the bright points of light from stars (plus the nearer planets and

Moon), what was seen at night was blackness, and for some time this seemed to demonstrate something rather strange about the universe. We are so used to this that it doesn't strike us as being odd. Black is what the night sky is – it just seems natural. Yet astronomers spotted something strange about that black night sky, something odd enough to be given a name of its own. It is referred to as Olbers' paradox.

Where are all the stars?

This problem was thought up as long ago as the 16th century, but it was most widely publicised by German astronomer Heinrich Olbers in the 1820s. It goes something like this. Let's assume that the universe is infinitely big – back then there was no evidence to suggest it wasn't. And we'll assume that stars are randomly scattered through space. Then if you pick any random line from Earth going up into the sky, you will eventually hit the surface of a star. Which should be bright. So, the whole sky should be a blaze of light, whichever way you look, rather than a black sheet with the occasional brilliant spot in it. The sky should be like one enormous star, day and night.

One of the first explanations that attempted to explain the blackness of the night sky was to suggest that there must be dark clouds of dust between us and many of the stars. Yet this was a problematic suggestion. The universe seems to have been around for a long time. Over the aeons, if it were constantly absorbing all that starlight, those particles of dust would heat up until they themselves begin to glow. There still should be light in all directions.

Another thought was that the universe was not infinitely big. We really don't know how big the universe is, but it is entirely possible that it has a finite scale with plenty of room for gaps between the stars. Take it to the extreme – imagine the universe only stretched as far as the stars we could see with the naked eye. The most distant thing visible would be the Andromeda galaxy. If that were the case, there would be lots of black sky.

This isn't a bad explanation, and it was one suggestion put forward by American writer Edgar Allen Poe in an essay in which he also gave the explanation that is largely accepted now (only to dismiss it). Poe said that if the universe were infinite, or just extremely large, then the only acceptable reason for the blackness is that there hasn't been enough time since the creation of the universe for the light to get to us. He argued that we have no reason for believing this to be the case, so instead, the universe can't be bigger than a size that allows for the gaps we see.

We now believe that his discarded speculation was correct. There is good evidence that the (visible) universe is 13.8 billion years old, so we can only ever see light that has been travelling for 13.8 billion years or less. Because light travels at a finite speed, any light that would take more than 13.8 billion years to reach us wouldn't have arrived yet. To fill in all those gaps in the night sky, we would need light to get to the Earth that set off far longer in the past.

It's big, but not *that* big

It might seem bizarre now, but as late as 1920 many astronomers thought that the Milky Way was the entire universe. For those who held that viewpoint, all the fuzzy patches in the sky, such as what we now know is the Andromeda galaxy, were nebulae within the Milky Way, similar to the Orion Nebula we visited earlier.

That the view of a Milky Way-sized universe was still acceptable by then is clear from what is sometimes known as the Great Debate, held at the Smithsonian Museum in Washington D.C.

on 26 April 1920. The contributors were American astronomers Harlow Shapley from the Mount Wilson Observatory, who supported the Milky Way as everything, and Heber Curtis from the Lick Observatory who thought that the Milky Way was just one galaxy of many in the universe.

One of the strongest arguments of the Milky-Way-as-universe was that if it were not true, then the Andromeda galaxy, for example, would have to be far bigger than it appeared to be. If that were the case, it would have to be much further away – and many astronomers could not accept that the universe was so big. It was, in a way, the astronomical equivalent of a dispute over the age of the universe, which had to be increased several times as earlier ideas simply couldn't accommodate new scientific discoveries.

The 1920 debate itself did not provide a widely accepted conclusion. This would take a few more years when American astronomer Edwin Hubble was able to measure the distance to the Andromeda galaxy, as we discovered when looking at variable stars and standard candles. Once Andromeda was shown to be far outside the bounds of the Milky Way, our home galaxy could hardly be the entire universe.

The cradle of … everything

As we have seen, galaxies and supermassive black holes seem to have evolved in an interactive process that still isn't wholly understood. But what we do know for certain is that everything you've ever seen* or touched, spoken to or met, is a product of

---

* In saying the Milky Way is everything you've ever seen, I mean directly, rather than as a result of technology, such as telescopes. Strictly speaking, even this isn't true as our nearest large neighbour, the Andromeda galaxy, is just about visible to the naked eye – but you need good eyesight and a clear enough sky that it's a rare experience in practice.

processes in the Milky Way galaxy. Our galaxy is far more than a location – it is the thing that made us.

We have already discovered how stars have created some of the elements heavier than hydrogen (with the exception of a spot of helium and lithium that also dates back to the big bang), and how supernovae and stellar collisions both assembled the elements heavier than iron and distributed that matter out into the galaxy to eventually form part of the Sun's protoplanetary disc. This would in turn give rise to the planets of our solar system, including Earth. The atoms that make you up had the same origin as the planet. Hydrogen came from the big bang, but the rest was distributed by exploding stars within the Milky Way.

Of course, this process isn't unique to our galaxy: it is happening in all of them. However, such is the scale of the universe that we can safely assume that, aside from hydrogen, pretty well all the atoms in your body were formed in and have always been part of the Milky Way.

Your atoms didn't come into existence when you were born, of course. They have been in other people, and in a wide range of animals and plants, bacteria and viruses. They were in dinosaurs and the earliest forms of life. All these things, and everything else on Earth (including the planet itself) are made up of atoms of just 92 types. We've managed to make a fair few more elements, reaching a total of 118 by 2005, though the heaviest of them are so unstable that they disappeared within a fraction of a second of their creation. However, only eleven are needed to make up 99.95 per cent of your body by weight.

There's more hydrogen than anything else, but because it is relatively light, your makeup by weight is around 65 per cent oxygen, 18 per cent carbon, 10.2 per cent hydrogen, 3.1 per cent nitrogen, 1.6 per cent calcium, 1.2 per cent phosphorous, 0.25 per cent each of potassium and sulphur and even less of sodium, chlorine and magnesium. There may be as many as 50 more elements present inside you – but only in tiny quantities.

It's not just that your atoms had been in other people, though. We can go further and safely say that you have atoms that were previously in the bodies of pretty much any historical character you could care to mention. It has been estimated that by the 2020s, there had been around 100 billion people alive, but you have around 100,000 times that many atoms in your body, so statistically it's very likely that atoms that have spent time in, say, Marie Curie or Isaac Newton will also have been in your body.

Of course, you are a tiny part of what makes up the Milky Way, but that same origin story could be applied to any other part of any other planet, orbiting any other star in our galaxy. The Milky Way is much more than an environment, it provides our basic building blocks.

A fledgling galaxy

In universal terms, most of the things that we experience are relatively young. Even our solar system, at around 4.6 billion years old, is a relative newcomer in a 13.8-billion-year-old universe. But the Milky Way as a structure dates back a lot further to around 13.5 billion years. Back then, our galaxy would have been very different.

Our own Sun is not unusual in being much younger than the galaxy taken as a whole. And many stars would already have to have lived and died to form the heavier atoms that we find in the solar system. But the earliest parts of what would become the Milky Way date back to the first 300 million years of the existence of the universe.

Along the way, as it gradually developed, the proto-Milky Way would have added to itself from the smaller galaxies that had formed in its neighbourhood – in the end, in space on a galactic scale, it's a gravitational 'who's better at pulling?' contest, with the winner able to rip matter from its competitors.

More than it seems?

According to many astrophysicists back in the early 21st century, what we see when we look out at the magnificent spectacle of the Milky Way is only a tiny fraction of the matter that is really there. Of course, there is plenty of matter in the galaxy that doesn't glow with its own light, from planets and black holes down to the scale of dust and gas. But even if we could see all that, the prediction was that we would be missing between five and six times as much matter of a different kind to the stuff that we can directly experience.

As we will discover, things proved to be rather different from the expectation of those figures from the early 21st century, but had it been a true picture, given an appropriate detector, you would see that the Milky Way was surrounded by a vast outer sphere of material known as dark matter, the existence of which had been predicted as far back as the 1930s.

Fritz Zwicky, the Swiss–American astronomer who we've already met as one of the leading lights in the development of theories on supernovae, spent some time observing a massive collection of galaxies known as the Coma Cluster. This is located around 320 million light years from Earth. Like most of the cosmic phenomena that we've visited, the Coma Cluster spins around – but when Zwicky made a measurement of the rate at which it was spinning and estimated the mass of material that was in it, he got a shock.

Admittedly, the estimate of mass was distinctly approximate – the cluster contains over 1,000 galaxies, which makes for a whole lot of stars and other matter to add up. However, within any acceptable range of estimate of the mass, Zwicky realised that the cluster should be splintering into pieces, because it was rotating far quicker than should have been possible.

It's not that galactic clusters have speed limits but a matter of simple physics. The faster something spins, the more force there is attempting to pull the components of it apart. Imagine putting a cup on a rotating disc, like one of the old vinyl albums on a record player. If the disc spins slowly, the cup will stay in place, held by the friction between the cup and the surface of the disc. But as the disc spins faster and faster, the effective centrifugal force* will result in the cup flying off. Similarly, if the Coma Cluster spun fast enough, the kinetic energy of the component stars and galaxies would overcome the force of gravity that holds the cluster together. And the cluster appeared to be spinning plenty fast enough for the whole thing to fly apart. Zwicky calculated that it would need 400 times the amount of material in it than he could account for if it were to remain stable. That's a dramatic difference and proved to be an over-estimate – but the Coma Cluster does appear to be revolving more quickly than it should.

Zwicky could only see one obvious reason for this: that the cluster had within it vast amounts of matter that did not glow in the way that stars and hot gases did. He called these

---

* Some pedants like to say that centrifugal force does not exist, and that in practice all the cup is trying to do is to continue to travel in a straight line, but it is a *centripetal* force towards the centre of the moving disc that is holding it back. However, this is simply one way of looking at what's happening. If you look at the setup from the cup's point of view, rather than from an external viewpoint, the disc isn't moving and there is indeed a centrifugal force, which is the easiest way to imagine what's happening. That's relativity for you: there is no absolute 'frame of reference' for understanding what's happening.

components *dunkle Materie* in German, which translates as 'dark matter'. And we must always remember that there is plenty of stuff out there in space that doesn't glow. But all the evidence is that the cosmos is dominated by the bright stuff. Remember 99.8 per cent of all the matter in our solar system is in the glowing Sun. The disparity of expected over observed mass should have been a flag for frantic investigation – but Zwicky's observations were largely forgotten until the 1970s, when American astronomer Vera Rubin, using a new type of electronic detector developed by her colleague Kent Ford, made far more detailed observations of the way that the components of individual galaxies rotated.

Galaxies with haloes

If a solid disc is rotating – one of the wheels of a car, for example – each part of the disc has to travel at its own specific speed, ranging from slowly near the hub to much faster at the rim. If you imagine the wheel undergoing a single rotation, in the time that takes, a point on the rim has to get all the way around the circumference of the wheel, while a point near the axle has a far shorter distance to cover – so the outer parts of the wheel travel much faster. But a galaxy isn't a solid disc. It is a collection of individual stars, each pulling on the others with the force of gravity.

When Rubin observed our near-neighbour major galaxy Andromeda, she discovered that stars near the edge of the galaxy were moving at similar speeds to those located near the centre. And the same thing is also true of our Milky Way. Just observing it now, we can't see this – it is far too big for the rotation to be visible simply by looking at the galactic spiral. Remember it takes around 210 million years for our solar system to take one turn around the galaxy. But Kent's technology enabled Rubin to detect the Doppler shift in the colours of the stars – and this made it possible to estimate how fast different stars were travelling. When we apply this technique to the Milky Way, the result is something like the graph that we're showing on the viewing wall.

View 38

Milky Way velocity of stars graph

To see outside the ship, use the QR code or visit www.interstellartours.co.uk/view38.html

The expectation from the laws of gravity were that there would be a 'rotation curve' showing speeds of stars that shot up as we move away from the centre of the Milky Way. These speeds would then gradually tail off from that initial peak – this expected rotation curve is shown in the solid line of the graph. But the reality is that the curve is much flatter, with less variation in speed and no tail clear off, as shown by the dotted line. Rubin suggested that this distribution of stellar speeds could be caused by a sphere of matter distributed around the galaxy that has become known as a halo (somewhat confusingly, given haloes are not spherical). But, looking at the Milky Way, we don't see all of this stuff floating around it. Like Zwicky, Rubin was hypothesising the existence of some kind of dark matter. Or more precisely, she was predicting the existence of invisible matter.

If the halo had been made of ordinary stuff, then you would expect that it would get in the way of the view of the rest of the universe. We might only ever be able to see the Milky Way in the night sky because of all that matter in the way. Yet, when we look outward from Earth, there is no obscuring halo. So, where Zwicky had been thinking of dark but ordinary matter, Rubin realised that this would have to be a different kind of substance made of particles that did not interact with electromagnetism – effectively totally transparent matter, but a substance that was still, confusingly, given the name of dark matter.

As Rubin and other astronomers pulled data together on a range of galaxies, such dark matter effects seemed to apply to practically all of the rotation curves. And like the Coma Cluster, these galaxies rotated too quickly – as Rubin's observations were refined, it was estimated that there was between five and six times as much dark matter present as there was expected to be normal matter. But what could this dark matter be made of?

Mysterious particles

The key to understanding dark matter was its lack of interaction with electromagnetism. Electromagnetism has an effect on all the familiar matter that the *Endurance*, the Earth and, for that matter, people are made from. We're not just talking about magnetic metals. It's the electromagnetic force that allows us to touch things, pick things up or to sit in chairs. If it weren't for the electromagnetic interaction between the atoms in your body and a chair, those atoms would slip past each other. In fact, when back on Earth, your atoms would be gravitationally attracted towards the centre of the planet (as would all the other atoms). No structure could exist. But there were already some particles known that did not respond to electromagnetic forces – particles known as neutrinos.

Neutrinos emerge from atomic nuclei when they undergo nuclear reactions. Their existence was hypothesised by Austrian physicist Wolfgang Pauli in 1930 – no one had ever detected a neutrino at this point, but energy was going missing from these reactions, and Pauli suggested it was carried away by a neutrally charged particle that he called a neutron. Just two years later, English physicist James Chadwick discovered what we now know as neutrons – particles that form part of the atomic nucleus. As a result, Italian physicists Edoardo Amaldi and Enrico Fermi suggested an alternative for the much smaller particle Pauli had dreamed up – the neutrino.

Neutrinos happily stream through matter as if it weren't there. On the *Endurance*, we don't get much neutrino flow except when near a star, but on the Earth, vast numbers of neutrinos pass through the planet – and everyone on it – all the time. It's been estimated that around 100 trillion neutrinos flow through your body every second without a problem on Earth.

The first detection of a neutrino was not until 1956, because they also tend to fly straight through most detection equipment. However, given that the Sun (and every other star) is producing so many neutrinos, a big enough detector will occasionally pick one up when it has a direct impact on an atomic nucleus. Neutrino detectors are typically large underground lakes of cleaning fluid. They are located down old mine shafts so the surrounding rock will cut out most other particles, with lead lining to keep out as many locally generated particles as possible. Some underground detectors go even further, using lead that was taken from the wrecks of old ships, as lead that was produced since the detonation of nuclear weapons will emit particles itself more frequently than older lead. When a direct impact occurs, there is a tiny flash of light that can be picked up by sensitive detectors lining the chamber.

We are not going to come across one out here in space, but it's worth taking a look at one of the historic neutrino sites, the Sudbury Neutrino Observatory in Canada on the viewing wall as they are quite spectacular in appearance: each of those circular nodules on the vast sphere is an individual detector.

View 39

Neutrino Observatory

To see outside the ship, use the QR code or
visit www.interstellartours.co.uk/view39.html

For some time, it was suspected that neutrinos had no mass at all, as is the case with photons, but they were discovered to have an odd behaviour called oscillation, where the particles switch between the three different types of neutrino when in flight. Theory required them to have some mass for this to happen. Not a lot, admittedly – neutrinos are about 4 million times less massive than the otherwise diminutive electron – but as long as they have some mass, they will have a gravitational effect (and even their energy contributes a little). There are so many neutrinos out there that it initially seemed feasible that they might have been the secret particles behind dark matter.

Unfortunately for neutrino fans, though, they failed to deliver. Over the years, we have become very good at detecting neutrinos, but none have been found acting as haloes around large bodies like galaxies. Worse still, dark matter has to be relatively slow moving if it is to have the desired effect of being an important part of the structure of galaxies. The now out-dated version of the big bang model popular in the early 21st century referred to 'cold dark matter', meaning that the matter was slow-moving, as the energy of particles is synonymous with temperature. For gravity to be able to pull matter into a clump to form a galaxy (or a dark matter halo), it has to have time to do so – not easy if the particles shoot past at near the speed of light. However, neutrinos usually do move at these extremely high speeds, meaning that they would evade capture as galaxies formed.

Although in the early days of dark matter theories it was thought possible that neutrinos could have been captured if early galaxies formed from vast super-galactic structures, evidence from the cosmic microwave background radiation has countered this. As we have seen, that radiation is light that had been flying around the universe since a few hundred thousand years after the big bang, and had such massive structures existed in the early universe, they would have been reflected in the pattern of the radiation's distribution.

MACHOs, WIMPs and axions

Researchers back in the early decades of dark matter research came up with a whole host of alternatives to neutrinos to be candidates for that missing mass in galaxies. These included those given the contrived names of MAssive Compact Halo Objects and Weakly Interacting Massive Particles – MACHOs and WIMPs*. MACHOs were simply ordinary matter that wasn't emitting electromagnetic radiation, just the kind of dark matter that Fritz Zwicky had assumed made up his *dunkle Materie* in the 1930s. This could be in any form, from tiny particles of dust up to black holes, as long as it didn't emit electromagnetic radiation – MACHOs are the stuff of the familiar face of the universe.

Unfortunately, though, there is plenty of evidence that there simply isn't that much ordinary matter out there. What's more, we know how ordinary matter acts in galaxies, and it doesn't involve forming these invisible haloes. MACHOs would be dark, but not invisible – they would interact electromagnetically and be relatively easy to detect. It was briefly suggested that there could be many undetected black holes in a galaxy that formed in the high-pressure environment of the early universe, rather than being produced by collapsing stars. Such 'primordial' black holes could have been any size, and some would have been less obvious than the traditional kind – but again, by the early 21st century it became clear that there was no good evidence for their existence.

WIMPs, by contrast, were supposed to be whole new particles, outside those understood to make up ordinary matter. These would be particles that did not interact electromagnetically but that had significantly more mass than a neutrino. Many detectors were built in the first decades of the 21st century to search for dark matter particles, but the WIMPs never showed up. Astronomers were faced with a difficult

---

* The terms MACHO and WIMP were astronomers' jokes from the late 20th century. The words, which have now dropped out of general usage, were used to describe two different kinds of person – pushy and aggressive (macho), or weak and passive (wimp). As these words are now obsolete, they may not get much of a laugh anymore – but then, astronomers' jokes rarely do.

decision – how long do you continue throwing expensive parties when no guests show up? Eventually the concept of WIMPs was abandoned.

Speculation alert

It is possible that WIMPs would have been found in the decade or so after 2023 – detectors continued to be built for some time during that period. However, the models of WIMPs that were favoured then predicted that they should have been detected with the equipment already available, and eventually budgetary restraints brought such a search to a close. It seems entirely likely that dark matter particles were never found. The description below of the rise of modified gravity theories is speculative, but these theories had been gaining increasing support in the 2020s.

A third option for a dark matter particle was a more speculative one called an axion. This sounds suspiciously like a brand name and was, bizarrely, named after a type of washing detergent that sounded suitably scientific. Axions were originally thought of (without any evidence for their existence) to explain an oddity in quantum physics, but had they existed, they would have had to have had extremely low masses. And as with WIMPs, nothing had been detected after nearly three decades of trying to find them by the 2020s, using devices that should have detected axions if they were truly there.

Tweaking Newton

As we have seen, at the end of the 20th century, most physicists and cosmologists were convinced that the effect described as dark matter was caused by some kind of extra stuff in the universe that had a gravitational effect but no electromagnetic interaction. However, as early as 1983, Romanian–Israeli astrophysicist Mordehai Milgrom had suggested an alternative reason to explain why galaxies were able to rotate faster than should have been possible.

Ever since Newton, it had been assumed that the force of gravity was universal – that it worked in the same way whatever the scale of object it was applied to. But this *was* an assumption, not a finding based on evidence. Although Einstein's work had resulted in modifications to the fine detail of gravitational attraction (and provided a mechanism for how it works in the warping of space and time by massive bodies), it did not challenge this assertion of universal applicability.

In science, it's arguable that all assumptions should be tested. It's certainly true that not everything acts the same way on different scales. As we have seen, a quantum particle travelling from A to B behaves very differently from the flight of a tennis ball. Milgrom suggested that on the scale of galaxies, gravitational attraction worked in a slightly different way from the scales at which it was conventionally measured – dealing with (relatively) small things such as planets and individual stars. His idea was known as Modified Newtonian Dynamics (MOND).

The MOND theory did well for a while, but supporters of dark matter particles were able to put forward an example where MOND struggled to provide an explanation in a structure known as the Bullet Cluster, an incredibly large-scale structure formed by two (or more) colliding clusters of galaxies, located over 3 billion light years from Earth. We can't visit it, but we can show you what it looks like, from a combination of X-ray and visual imagery.

View 40

The Bullet Cluster

To see outside the ship, use the QR code or
visit www.interstellartours.co.uk/view40.html

The cluster has an unusual shape with a central, vaguely bullet-shaped blob and two bulbous outer regions, which is relatively easily explained if there is WIMP-based dark matter. The idea is that the ordinary matter largely stayed in the middle of the resultant cluster after the collision, limited by pressure from high-energy radiation. But the dark matter, which was not influenced by electromagnetism, passed straight through the middle until slowed by gravity to form those bulbous lobes, subsequently pulling enough material out of the centre to be visible.

If galaxies were purely ordinary matter, influenced by modified gravity, it's less easy to see how this structure could be formed. For the dark matter enthusiasts, this one example (and a few other similar issues) meant that MOND could instantly be dismissed. And that remained the dominant attitude for a couple of decades, with the relatively small number of MOND supporters sometimes shouted down at conferences. But by the 2020s, there was growing resistance to this attitude. This is because far more galaxies had properties that could easily be explained by MOND – and were difficult to explain using dark matter – than there were galaxies that proved a problem for MOND.

Science may attempt to be objective and emotionless, but scientists are people and often get so caught up in a theory that they resist change. For a long time, dark matter particle supporters continued to reject enhanced variants of MOND. As Professor Stacy McGaugh, a former dark matter supporter who realised that sticking rigidly to this viewpoint whatever the evidence was not scientific, pointed out, the real problem with dark matter without a successful candidate for what the particles were is that it can be tweaked to explain anything, and as such, lacks effective predictive ability – an essential for a good scientific theory.

Professor McGaugh commented: 'Dark matter, being invisible, allows us lots of freedom to cook up an explanation for pretty much anything. My long-standing concern for the dark matter paradigm is not the failure of any particular prediction, but that, like epicycles, it has too *much* explanatory power. We could use it to explain pretty much anything. Rotation curves flat when they should be falling? Add some dark matter. No such need? No dark matter. Rising rotation curves? Sure, we could explain that too: add more dark matter. Only we don't, because that situation doesn't arise in nature. But we could if we had to.'

By being non-specific about dark matter particles, supporters could explain a vast range of possible outcomes in the way that galaxies behaved, which inevitably included (with much tweaking) what was observed. By contrast, MOND produces very specific predictions, which most galaxies happen to fit. Some have suggested that there is *some* dark matter, producing the anomalous effects, but that mostly what we are seeing is a modified gravitational effect in galaxies.

Eventually, the evidence against dark matter particles producing most of the effects ascribed to it was too strong, and although a fully satisfactory theory has yet to be clarified, by the time the *Endurance* was commissioned, there was no serious support for dark matter particles being the cause of the majority of these phenomena.

Going large

Having taken in the whole of our galaxy, we are going to go one step further. Despite having said that we can't travel further than 100,000 light years, we are going to take a brief excursion to what could be considered the centre of the universe – the point where the big bang occurred. And unlike every other trip we've made so far, this journey will be instantaneous.

THE BIG PICTURE 10

Where it all happened

The reason we were able to get to the location of the big bang instantly from our position 100,000 light years outside the Milky Way is that we haven't moved. Remarkably, the exact point from which you can see the Milky Way from the outside is also where the big bang occurred – the whole of the visible universe expanded from this point.

This sounds like an unlikely coincidence. Why should the location of the big bang happen to be here, so close to our galaxy in universal terms? The reason is simple. We could have made the same claim anywhere. For example, if you were still back on Earth, wherever you were located it would be there. Perhaps you will read *Interstellar Tours* to revisit everything you've experienced on the tour, in your armchair at home. You will still be sitting at the point where the big bang occurred. Wherever you are in the observable universe, you are at the location of the big bang.

Think of the universe as a bubble that was blown from an infinitesimally small piece of material. Every point that is part of the bubble was originally in the same location, the place where our bubble big bang took place. It doesn't matter where we are located in the material of the bubble, all of it came from the same place. The same view applies to the expanding universe. We can't point to a centre and say, 'That's where it came from,' because the universe didn't expand from a point in space – it was space itself that expanded.

Before we take our final trip of the tour, let's take a few minutes to put our understanding of the universe into context. It's not a place that we can visit, like a black hole or a galaxy. It is, by definition, everything. Arguably, understanding the universe is something that sits between philosophy and science, so it makes sense to get a historical view to put it into context.

The mythical view

From the earliest times, creation myths have been written to explain where the 'everything' that is the universe came from. Human beings are born storytellers. It's not our natural way just to pass on a series of dry facts, as usually happens in a science lesson. Stories are more appealing, more memorable and a fundamental part of civilisation. Creation myths are, in this sense, storytelling, not science. It's important to understand, though, that by calling these stories about the origin of the universe 'myths', we are not insulting them, nor those who consider them sacred.

In modern usage, if we say, 'That's just a myth,' in response to something, it suggests that we are not particularly impressed with the idea. When this is applied to something like an urban myth, the term describes a commonly held belief that isn't true. But the ancient myths were something entirely different. A myth was a story with a point. It provided a response to a deep question about our existence by reference to an exotic setting, usually distant in time. A myth was never intended to be the same thing as history or science – it didn't purport to be fact-based truth, but rather was a way of giving an understanding, a feel for the nature of today's reality, through a story about the past.

The early tellers of creation myths didn't have in mind the vast expanse of what we now think of as the universe. They could never have envisaged anything on that scale because all they had experienced was a very local landscape. Their universe was the Earth and the heavens (a rather vague term for everything that wasn't on the ground). The sky was effectively a roof to the land and the sea. Up in that sky were placed the Sun, the Moon and the stars – but they were never envisaged as the kind of vast objects we now know them to be. Outside of the universe, most creation myths placed (as the word 'creation' suggests) a creator, who was given the role of the guiding hand behind the design of the universe.

Danger, god(s) at work

Many scientists are suspicious of the idea of a creator, but it appeals to common sense based on experience*, and it's a viewpoint that many individuals still hold. William Paley, a Victorian priest, famously argued that the complexity of life implies the existence of a creator. He used the example of finding a watch, lying on the ground. No one would mistake the watch for something that was part of the landscape, like a rock. It seemed too complicated and functional. A watch has obviously been 'manufactured' (which literally means 'made by [human] hand').

---

* Note that in saying the existence of a creator appeals to common sense based on experience does not necessarily mean it is correct. Plenty of common-sense ideas are wrong. But equally, there is nothing in science that can disprove the concept of a creator.

Paley was drawing a parallel with the complexity and function-ality of living things, but those inspired to produce creation myths surely had similar thoughts about the world and sky above them.

One of the earlier creation myths we know in detail is the Ancient Egyptian description of the origins of the universe. To us, Egyptian myths can be confusing. Not only were there a multitude of versions, but also there was no book or myth that was considered definitive. For most of history, they did not restrict themselves to having a single god for a particular feature of nature (a single Sun god, for example), and any particular god could have a number of aspects, where they looked, behaved and were even named differently. So, for example, the Sun was considered at various times to be a number of individual gods or was an aspect of a collection of gods.

Most versions of the first steps of creation in Egyptian myths came from water – something that is also seen in the Book of Genesis in the Bible. This seems to reflect the importance of the River Nile to Egyptian civilisation. The mythical aborig-inal waters, called Nu or Nun, divided to provide dry land, on which was revealed the first of the gods, Atum. He spat out Shu, the god of the air and Tefnut, the goddess of moisture whose daughter was Nut, the goddess of the sky, and whose son was Geb, the Earth god.

It was these two siblings whose children became the most important gods in the Ancient Egyptian mythos, Osiris, Isis, Set and Nephthys (although these were just a fraction of the full pantheon). In some variants of the creation myth, originating in the lower kingdom, the first of the gods was Ra, the Sun god, who was transformed into the Sun's disc, the Aten. In one short period of Egyptian history, this disc was thought of as the only god in the monotheistic religion of Akhenaten.

Early Chinese writings reflect the other common mythical starting point for the universe – as an egg. In these myths, the creator god, P'an Ku emerged from an egg that seems to have just been sitting around since the beginning of time without ever being created. In the process of his hatching, the two halves of the egg became the sky and the Earth. But in this story, the true act of creation was one of sacrifice. At the age of 18,000, on his death, P'an Ku became everything that filled that egg-formed universe. His flesh became the soil, and his blood formed the rivers. We, it seems, are descended from his fleas. In this myth, god and creation are all the same thing.

Probably the best-known creation myth in the West, though, come in the first pages of Genesis, the opening book of the Bible, which is a story that was developed from earlier Babylonian creation myths. The spirit of God moves over the face of the waters (we aren't told where these waters came from, nor what is supporting them) and creates the heaven and the Earth. To these are added light, plants, the heavenly bodies and living things with man coming last.

Although many of the ideas in the Genesis creation myths go back much further, it probably came into its present form about the same time as another group of myths were being formulated – those of the Ancient Greeks. The Greek universe began with the void, emptiness – although confusingly, they also referred to it as chaos, which seems to imply some contents, as you can't have chaotic nothing. The earliest god was usually Eros, born from the golden egg of the bird Nyx, an egg that split into two to form the sky and the Earth, similarly to the earlier Chinese myths, although this version probably evolved independently. The idea of everything starting from an egg is a natural one – and when you see the universe as Earth and sky, it's not a great stretch to see these as two halves of a great egg.

In some versions – like the Egyptians, the Greeks had no definitive text – a mother goddess Eurynome was there alongside Eros and actively brought order out of the chaos. But the Greek gods we are probably most familiar with – from Zeus onwards (who himself formed the template for the Roman god Jupiter), were the children of the Titans, a first generation of gods whose parents were Ouranos, the sky, and Gaia, the Earth.

The Greeks had their myths, like every other civilisation of the period, but they also went further. Rather than be satisfied with what could be called a magical explanation (the universe works because a god or gods makes it work), they looked for rational explanations of the detail of how the universe functioned.

Early science

Probably the first 'scientific' cosmology – a self-consistent picture of the universe and its origins that was built on physical forces and structures, rather than the whim of the gods – was the work of an early Greek philosopher, Anaximander, working in the first half of the 6th century BC. Anaximander's aim was not to displace the gods but simply to explain what was observed.

As one obvious observation was the presence of fire in the sky (pretty much the only known source of light on Earth then), Anaximander suggested that the universe was full of fire, initially as a chaotic whole and later hosting a vast, shell with holes in it. (There was no scientific explanation for its existence.) As this shell had holes in it, the firelight shone through, producing the lights in the sky – the Sun, Moon and stars – and the easily felt heat that emanates from the Sun.

At the time, limited thought was given to any internal structure, but this was developed over the years, before becoming fixed in the form described by Aristotle in the 4th century BC. The model that Aristotle and his contemporaries devised would stay in place for over 2,000 years. (Compare that with our current picture of the universe, which has lasted maybe a tenth of that.) Aristotle's ideas would be tweaked to fit observation better, but the core of the model remained the same.

Not surprisingly, the starting point was the Earth. Nowadays, with our eyes firmly fixed on the big picture, putting the Earth at the centre of things seems egocentric and naïve. However, in doing this, Aristotle had a perfectly good (if incorrect) scientific argument. He believed that matter was made up of four different elements – fire, air, water and earth. The first two were light and had a natural tendency to move away from the centre of the universe, a tendency known as levity. The other two were heavy with a tendency to seek the centre of the universe – so-called gravity. For this idea to work, the Earth had to be at the centre of things or all the heavy stuff would head for somewhere else.

Around the Earth, Aristotle's lined up a series of crystal spheres, one inside the other, Matryoshka doll style. Working outwards, first came the sphere of the Moon, then one hosting Venus, then Mercury, the Sun, Mars, Jupiter and Saturn. Finally, almost on the outside of the universe, came the sphere of the fixed stars. By saying these stars were fixed (a term that is still used), there is no suggestion that they stay in the same place in the sky. Instead, they were fixed in the sense that they all moved together on the same sphere, while the planets (originally meaning 'wandering stars') moved against them.

There was a kind of mechanism that meant that each sphere was driven by the sphere outside it. This was arguably part way between science and myth, so the exact nature of that mechanism didn't have to be fully explained. When it came to the outer sphere, there was a bit of problem. Aristotle's physics required that something was pushed in order to keep it moving. To deal with this, the sphere of the fixed stars was put in the hands of a deity known as the prime mover. Despite this, though, Aristotle's was a primarily scientific cosmology. Once it had that divine power, everything else became a sort of heavenly clockwork*.

Aristotle did away with the sea of fire, making the Sun the source of all light. Not only did he think, as is the case, that the Moon and planets were lit by sunlight, he also thought the stars only provided reflected light. This was potentially a problem as the Earth should cast a shadow on the sphere of the stars, eclipsing many of the stars. The work-around Aristotle devised was to suggest that the Earth's shadow only stretched as far as Mercury, leaving the stars untouched.

This was a picture of a universe that was about the same scale as the solar system. Enormous by the standards of Ancient Greek travel, yet still relatively limited. Aristotle did not specify how big that universe was, but about 100 years later, the outstanding mathematician and engineer Archimedes would have a go at estimating the size, as part of the apparently pointless exercise of estimating the number of grains of sand the universe could hold.

---

* Although the Ancient Greeks did not have mechanical clocks, they did have mechanical devices and within a couple of hundred years of Aristotle's life had produced the so-called Antikythera mechanism, a mechanical orrery with a geared structure that provided a direct mechanical model of some of the movements of the heavens.

The Sand Reckoner

Archimedes is probably best known for leaping out of a bath, shouting, 'Eureka!' when he had an inspiration about the way that objects displaced liquids. But both his mathematics, which came close to an early form of calculus, and his mechanical creativity were remarkable for the time. His book, *The Sand Reckoner*, was anything but a practical exercise, though.

The idea was to show up the limitation of the Greek number system of the time, which only effectively went up to a myriad, which was 10,000. You could have a myriad myriads (100,000,000), but anything more than that was not considered. To show it was possible to go far beyond this figure, Aristotle invented an extension to the number system that could deal with immense values. To be honest, it was never a practical approach but was rather a means to prove a point.

This number system was put to use in envisaging filling the universe with sand and counting the grains. Greek maths primarily depended on geometry, and Archimedes was able to use this, along with a few basic assumptions (such as the Sun being bigger than the Earth, and the Earth bigger than the Moon) to give a distance across the universe of around 10 billion stades. These are units based on the size of a stadium, just as we often use football pitches in estimations now. Each of the stades was around 180 metres (590 feet) in length*, making his universe 1,800 million kilometres (1,120 million miles) from side to side.

Saturn's orbit is 2,800 million kilometres across, so this (impressively) is the right order of magnitude for a guess of the size of the then known universe. Even better, Aristotle tried

---

* The size of stades as 180 metres is a bit of a guess, as not every stadium had the same length of running track, so the unit varied from city to city.

out a different structure for the universe. An astronomer called Aristarchus had written about a possibility that the Sun, rather than the Earth, was at the centre of things. We don't have that book – Archimedes' use of it is our only tantalising reference to its existence. But with that form, Archimedes reckoned the universe would have to be about 10,000 times bigger. That's a size that would encompass the whole solar system. Translating from his odd units, Aristotle calculated it would take 10^{51} grains of sand to fill the ordinary universe and 10^{63} grains to fill the Sun-centred version.

The spheres that Aristotle envisaged making up the universe had a serious problem. This was something known observationally at the time, but Aristotle was at heart a philosopher, rather than a true scientist. His philosophy infamously required women to have fewer teeth than men, but he didn't bother to count them to see if it was true. Yet you only had to plot out the path of Mars in the sky from night to night to realise that there was something wrong with the crystal spheres.

Scientific thinking replaces philosophy

As we saw when looking at the problems that novae originally caused, if you plot the path that Mars takes through the sky, based on Aristotle's picture, you would expect it to follow a continuous path, making a circle around the central body of Earth as it rotates on its crystal sphere. Instead, Mars suddenly reverses its direction in a process known as retrograde motion. It effectively performs a loop the loop in the sky, which was not a practical possibility if it were fixed in space on its sphere.

This strange motion is inevitable because Mars and the Earth are both rotating around the Sun, each travelling at different speeds on orbits that aren't concentric circles. So, as seen from Earth, the orbit of Mars will seem to loop back on itself as the faster moving Earth overtakes it – but this wasn't a possible explanation when using Aristotle's model of the skies. Soon after it was first formulated, Aristotle's universe had to be modified, adding complexity. Instead of being fixed to the main crystal sphere of the Mars orbit, the planet was thought to be attached to a second, smaller sphere called an epicycle. This was embedded in the main sphere. As that rotated, so did Mars' mini-sphere, an effect that gives us the expression 'wheels within wheels'. Make the size and sphere of the smaller Mars sphere right and it would produce the exact required loopy motion. The same approach was taken for other outer planets.

The version of this spheres-and-epicycles model described by the late Greek astronomer Claudius Ptolemy in the 2nd century AD would remain the definitive description of the universal workings until the 16th century. First Polish astronomer Nicolaus Copernicus (aka Mikołaj Kopernik) came up with a Sun-centred, Aristarchus-like model in 1530, published on his deathbed in 1543. Later that century, Danish astronomer Tycho Brahe produced an arguably more acceptable variant that would still work with Aristotle's physics, where the Earth remained at the centre of things, but only the Sun orbited the Earth, while everything else moved around the Sun*. And later, Italian polymath Galileo, born in 1564, became the most famous supporter of the Copernican model.

Famously put on trial for his undiplomatic approach to supporting Copernicus' system, Galileo added new observations

---

* In a sense, this is what does actually happen, at least as seen from the surface of the Earth. It's all down to your viewpoint. Relativity again.

with a telescope to help dismiss the Ancient Greek concept. Galileo didn't invent the telescope, but he did make some early instruments, and with them he was able to spot detail not previously available. He discovered that there were four moons orbiting around Jupiter. Here was direct evidence that everything did not rotate around the Earth, calling the Ptolemaic system into question.

As evidence grew, the Copernican model of the universe (or Tycho's, for that matter) made too much sense to be rejected, throwing away the need for the arbitrary complexity of epicycles. From the 17th century onwards, the picture of the universe was very similar to our current view of the solar system. At the centre was the Sun. Then came Mercury, Venus, the Earth (with the Moon orbiting it), Mars, Jupiter and Saturn. We may have added a couple of extra planets since, but in essence, it had the structure we now know to be true.

The stars were no longer thought to be on a crystal sphere, although this brought forward new questions to be answered. How did the stars manage to stay in place? And if the planets were just hanging in space, what kept them rotating around the Sun? Isaac Newton provided two essentials to deal with these questions. Firstly, he threw away the Ancient Greek idea that things needed a push to keep moving. His first law of motion made it clear that things in motion would keep moving unless something acted to stop them.

Secondly, Newton brought in a new version of gravity, representing it as a strange force that acted at a distance to keep the planets (and us) in place – although it would take Albert Einstein to come up with an acceptable explanation of how gravity works. This is where the prehistory of science catches up with what we've already seen on our tour when it comes to the action of gravity and the development of Newton's theories to bring them into the general theory of relativity.

It's growing

As we discovered when looking at quasars, the distance to the edge of the observable universe is now much greater than the 13.8 billion light years we might imagine from the age of the universe. This is because the universe is expanding, something that was first realised when it was noticed in the first part of the 20th century, notably by American astronomer Edwin Hubble, that almost all galaxies are red shifted. This reduction in the energy of light, shifting light colours in this direction on the spectrum, means that the other galaxies are moving away from us.

There are a few anomalies, notably the Andromeda galaxy, which are blue shifted. This is because these examples are so close (relatively) to our galaxy that they are moving towards us. But otherwise, everything moves away. Going back to our bubble model of the universe, if we imagine the bubble was a partly blown-up balloon and drew some dots on it to represent galaxies, then blew the balloon up a bit more, we could pick any dot at random, and all the rest of the dots would be moving away from it. As we saw, this is not because that dot is the centre of the universe, but because space itself (represented by the balloon) is expanding.

The difficulty in envisaging this is because the surface of the balloon is two-dimensional, expanding into three-dimensional space, but the expanding universe is already three-dimensional. However, given such a model and the evidence for expansion, it wasn't long before running the model backwards produced the concept of the big bang (although the theory has a more sophisticated mathematical basis). If we run our picture of the expanding universe in reverse, we reach a point where everything was in one place, at the point in time where everything started expanding. This is the big bang.

The big bang theory describes a universe that began around 13.8 billion years ago with the entire universe in the form of an infinitely small point called a singularity. At least, this is one way of looking at it, as there is some dispute over exactly what the now-familiar term 'big bang' means.

We know where the name itself came from – ironically, coined by a strong opponent of the big bang theory, English astrophysicist, Fred Hoyle. Along with Thomas Gold and Hermann Bondi, Hoyle had come up with an alternative theory known as steady state, inspired by a trip to the cinema in Cambridge to see the anthology supernatural film, *Dead of Night*. One of the innovative things about this movie is the way that the last scene leads straight into the opening one – remove the credits and it is a circular process with no beginning or end.

Hoyle and his colleagues came up with a model where the universe expanded, but matter was always being created, giving a universe that had no beginning or end. Hoyle was an excellent science communicator, rare among the academics of his day, and he gave a radio broadcast on the idea, commenting that the steady state theory 'replaces a hypothesis that lies concealed in the older theories, which assume, as I have already said, that the whole of the matter in the universe was created in one big bang at a particular time in the remote past'.*

From this, it seems that Hoyle had it in mind that the 'big bang' was the moment of creation, although since his broadcast, some physicists have applied it to a point in time a tiny fraction of a second later, when the universe began to expand. A lot of people wonder, 'What came before the big bang?' and in the basic big bang model, the answer is very simple. Nothing came

---

* It has often been said that Hoyle used 'big bang' in a derogatory way in the broadcast, but he claimed that it was simply to give the other theory a neat label, as they had done for steady state.

before the big bang. Not only was there nothing before, there was no before, because it was the beginning for both time and space.

Something Hoyle pointed out in his radio talk was that in one sense the big bang theory isn't fit for purpose. He commented: 'On scientific grounds this big bang hypothesis is much the less palatable of the two. For it is an irrational process that cannot be described in scientific terms ... it puts the *basic* assumption out of sight where it can never be challenged by a direct appeal to observation.' (Emphasis in the original.)

The big bang gives us no more scientific explanation for the start of everything than did those creation myths. It gives us a beginning in a hypothetical singularity (a concept that breaks the known laws of physics), but it tells us nothing about where those laws come from and *how* the universe came into being. Nonetheless, it remains the best current theory. Although the big bang is perhaps treated more sceptically now than it was in the 21st century, there is as yet no useful replacement for it.

We may never have a good scientific theory for the origin of the universe that is supported by scientific evidence. You will see many science books giving a detailed description of the process from the big bang through the various early stages of development of the universe, and much of this makes sense scientifically, but it is detail that doesn't help us with where the universe came from in the first place. Pick the creation myth of your choice.

From big picture to small detail

After taking a moment's pause at our most remote location within the universe to look into the often philosophical and historical big picture of the universe as a whole, we are heading back towards Earth. But the tour is not quite finished. Not only can we explore the spacetime oddities that arise from travelling faster than light, but we will also have a chance to take in some of the iconic sights of our own solar system.

It might seem that after viewing some of the most exciting aspects of the galaxy that the planets, which have been familiar celestial companions for thousands of years, would turn out to be uninspiring. Yet, there is always something special about coming home from a long voyage and seeing familiar sights there with different eyes – and this is never more the case than when we are returning to the only known location of intelligent life in the universe.

TIME, LIFE AND OUR BACKYARD

11

Talking to the past

In taking our trip outside the galaxy we have travelled the furthest distance we will cover on this voyage. To get a good view of the Milky Way, we had to reach around 100,000 light years from home. Thanks to the hyperspace drive, we can achieve this limiting distance (but no more) in the characteristic 42 minutes. However, as we head back towards the solar system, it's worth thinking what the implications of the ability to travel such distances have on the flow of time – and for this, we have to revisit Albert Einstein's work.

When we make a jump through hyperspace, we can arrive at our destination far faster than light would take to make the same journey. After all, the journey that is currently taking us 42 minutes to cover would take light 100,000 years to traverse. This distinction becomes interesting because, with certain exceptions, the ability to travel faster than light enables messages to be sent backwards in time. And once that's possible, the familiar fabric of reality suffers an irreversible shock.

When considering what message to send back in time, a common first thought is that this would be a fool-proof way to win a lottery – or of making money by any other means where knowledge of the future would give a distinct advantage, for example, when playing the stock market. Admittedly, such an advantage would not last long, as those organising lotteries and stock markets would rapidly realise the risk and these familiar forms of gambling would have to be abandoned. However, we shouldn't assume there wouldn't be a brief opportunity arising for those who would like to cheat.

This has already happened when a less dramatic form of time travel was introduced in the 19th century. At a time when communications took days to get from one end of the country to the other, bookmakers would take bets on horse races in England well after the race had been run. When telegraphs, sending messages along electrical wires, were first introduced, there was a sudden outbreak of unexpected winners among those placing bets, because it was possible to find out the results from a distant track while it was still possible to place bets on the outcome.

The telegraph companies were persuaded not to transmit any more messages with racing results, but this simply led to coded messages being used. A well-documented example involved a race at the Epsom Downs track, about twenty miles from London. A telegraph message was sent to Epsom, apparently innocently asking a friend if they would send some luggage and a shawl to the capital. The reply came: 'Your luggage and tartan will be safe by the next train.' Tartan proved to be a code word that identified a race winner. Not long after, the bookies gave in to the shift in the timing of messages and stopped taking bets once a race had begun.

Appealing though becoming a lottery or horse-race winner may be, the ability to get rich quick just scratches the surface of what can emerge when information can be sent backward in time, because this process threatens to disrupt the nature of causality itself. It is generally assumed that for event A to cause event B, event A has to occur earlier in time – but this is no longer necessary given a suitable time travel mechanism.

Even a small displacement back in time puts causality into a spin. Imagine I had a time phone that could receive a message from just one second in the future. Like all good modern pieces of equipment, the phone can be switched on and off remotely. With the time phone on, I send a message back in time to switch it off, a second previous to my sending the message. As a result, the phone is off in the past. But that means that when I try to send my message back, it won't be possible to send it, because the phone is switched off. Only, if the message wasn't received, then the phone would have still been switched on. And so on, in an infinite loop.

Extend the time duration for sending back a message further and things get even more mind-boggling. Think of your favourite book or piece of music. There was some point in the past when it was first written. But what if we use our time phone to send back the text of the book, or the sheet music, to a time before it had been written. Now the person who 'wrote' it, could simply copy their original down from the future version and publish it. They would not ever have to think it up in the first place. Imagine we did this. Then, where did the book or piece of music come from? It wasn't written in the past, just copied out. It wasn't written in the future, just sent back. It has somehow emerged from nowhere.

It's all relative

Of course, these remarkable causality-twisting outcomes depend on an ability to send a message into the past – but why does faster than light travel make this possible, and would it be possible with the technology used to take the *Endurance* on its interstellar journeys?

As we saw when looking at the necessity to travel faster than light to reach the stars in a practical timeframe, we are prevented from accelerating through the light speed barrier by Einstein's special theory of relativity. And one of the implications of that theory is that time slows down when travelling at high speed. But what does that really mean?

A starting point to understanding this is just digging into the meaning of the word 'relativity'. We tend to associate relativity with Einstein, but it was Galileo who formulated basic relativity. What he pointed out was that we can't say arbitrarily whether something is moving or not. We have to note what it is moving with respect to. By default, on our home planet, that something is the Earth itself. But that isn't always the best 'frame of reference' as physicists refer to this concept.

Galileo is said to have demonstrated the importance of relativity in a boat, being rowed across Lake Piediluco in Italy. You can see this beautiful setting on our viewing wall.

View 41

Lake Piediluco

To see outside the ship, use the QR code or visit www.interstellartours.co.uk/view41.html

According to a possibly apocryphal story, Galileo asked one of his friends, Stelluti, who was with him in the fast-moving boat, if he had something heavy and was handed Stelluti's house key. Galileo threw the heavy iron key straight up into the air as hard as he could. Now, the boat was moving quickly. So, Stelluti reasoned, by the time the key fell back down, the boat would have moved on and his key would sink irretrievably into the deep lake. He dived into the water behind the boat to try to catch it, just in time to miss the key falling back into Galileo's lap.

The lesson here for the soggy Stelluti was that the boat was moving with respect to the lake and the surrounding land. It was moving in the frame of reference of the Earth's surface. But as far as the key (and the people in the boat) were concerned, the boat wasn't moving. In the key's frame of reference, the boat was stationary. Galileo would later point out that in a steadily moving boat with no windows you could not do any experiment that would detect that motion – because from the viewpoint of the experimenter it was actually the water that was moving backwards, rather than the boat moving forwards.

Note, the essential requirement for steady motion, by the way. If the boat accelerates, we can detect it, because a force is being applied and we feel the result of that force. Think of being in a car, or a plane on the runway, as it accelerates. Even without looking out of the window you are aware of the motion as you are pressed back into your seat.

The same applies to a spaceship as seen from Earth. No matter how fast the spaceship is moving from a person on the ground's viewpoint, if the spaceship is moving at a steady speed, to people onboard that vessel it is not moving, and no experiments that they undertake will show that it is.

Slowing time

With the basics of relativity in place, we can take on the extension of it provided by Einstein's special theory. The easiest way to imagine the 'time dilation' effect that arises from special relativity is by using an imaginary device called a light clock*. This is a clock where a beam of light bounces up and down between two mirrors, acting as the regular ticking of the clock. We imagine such a clock on a spaceship that we can somehow see into from the Earth. When the spaceship is stationary in space compared to the Earth, then viewers on Earth will see the light beam moving vertically up and down, just as passengers on the spaceship do.

But now imagine that the spaceship is moving at a constant high speed away from the Earth. The passenger on board will see no difference to the light clock. The light will still travel directly up and down, because in the passenger's frame of reference the clock is not moving. But the observer on Earth will see the light beam travel at an angle, as the bottom mirror will already have moved further away from Earth by the time the light hits it, so the light would be seen to travel on a zig-zag path.

For most moving things, this observation is entirely consistent with Galilean relativity – but there's a catch when it comes to light, which is that, as we have already seen, its speed is constant in a particular medium, no matter how fast you move with respect to it. By combining this fact with the requirement for the light to travel further from mirror to mirror as seen from Earth, the special theory of relativity makes it clear that time on the ship will slow down as seen from Earth. If light is travelling at the same speed yet covering a greater distance, this can only be achieved as time is progressing more slowly.

Confusingly, this is a symmetrical effect. If the passengers on the ship looked back at the Earth, they would see time running slowly there. This is because from their viewpoint, the ship is

---

* The light clock is only described here, but the maths that shows how it works is simple and is presented in the appendix of *Interstellar Tours*.

not moving; instead, Earth is travelling away from the ship. However, this symmetry is broken when the spaceship has a force applied to it to accelerate, which the Earth doesn't experience. As a result, when the spaceship returns to the Earth it will arrive in the Earth's future, because Earth's time will have ticked by faster than the time on the ship. The faster and longer the ship flies, the further into the future it will have travelled when it gets back home.

Moving into the future seems back to front with our assertion that travelling at ultra-high speeds should make it possible to send a message back in time. To do this, we need an extra ability. As well as a fast-moving spaceship, we need a message that travels faster than light.

Calling the past

Let's imagine that we had a magic device that could instantaneously transmit a message from A to B, breaking the light speed barrier. That would be a handy thing to have anyway. Even on Earth, message delays of a fraction of a second can still cause problems with high-speed electronic transactions, such as those used for stock trading. But communicating with, say, the Mars colony would benefit hugely from instant messaging.

Depending on the position of the Earth and Mars in their respective orbits, it can take between five and twenty minutes for a radio signal to cross space from planet to planet. Before there was a human presence on the planet, this meant that it was impossible to use effective remote-controlled Martian rovers, limiting what was possible to semi-autonomous (and hence limited) exploration devices. And once humans were there, it made communication a slow business. Waiting 40 minutes for the response to a question results in a call to Mars being distinctly tedious.

Even the hyperdrive doesn't help here – with its 42-minute journey time, we can't send messages faster than light over such a (relatively) short distance. But if we had an instantaneous communicator, interacting with the Martian colony would be far smoother.

So far, so good. But the value of an instant communicator becomes far greater if we can combine it with a high-speed relay station. Let's imagine we send out an unmanned spaceship from Earth and get it up to a high percentage of the speed of light. After a few years, onboard time as seen from Earth would be, say, a year behind. If we could get an instantaneous message to the ship, then that message would have travelled back in time.

Of itself, this isn't particularly useful, because anything we could send would only really have relevance on the Earth. As far as the ship was concerned, this might be a message from the future, which could be intrinsically interesting, but not of great practical value. However, bear in mind that the effect is symmetrical. From the ship, it is the Earth's time that would have run slow. So, if the ship could send an instantaneous message back to Earth, it should arrive at Earth two years before the original message left Earth, making anything from cheating the lottery to twisted causality available.

In practice, there is no such thing as an instantaneous transmitter. It is not physically possible. Although there is a quantum effect called entanglement that does appear to communicate at any distance instantly, it can only send random information and cannot pass on a message. However, hyperspace drive does enable us to send a message across any distance (up to around 100,000 light years) in just 42 minutes. We frequently get asked if this means we can send a message back through time. But the answer is no.

A ship travelling through hyperspace is not moving through spacetime in any normal sense, and so does not itself experience any time dilation. A ship like the *Endurance* could not act as a high-speed relay ship, because onboard time is not dilated. But equally, it does not provide a near-instantaneous transmission to a conventional relay vessel that *has* experienced time dilation because just as acceleration effectively resets the clocks, so does travelling through hyperspace. There is still no way to send a message back in time.

For that matter, when you arrive back on Earth, exactly the same amount of time will have elapsed there as has passed by on the ship. You won't arrive back in the far future, as you would if making the journeys that we have at near light speed. This is good to know, as otherwise by the time you returned all your friends and family would be dead and the civilisation and culture you knew would be likely gone. Humanity might even have ceased to exist.

The pale blue dot

As we re-enter the solar system, we are going to drop out of hyperspace in a moment at what might seem a strange location – we will be well outside the orbit of Neptune. But what we're going to do is recreate a famous image from the past. Right at the start of our voyage we encountered Voyager 1, the probe from the 1970s with a distinctive gold disc onboard.

At our current position, we are significantly closer to the solar system than Voyager 1 was when we caught up with it. To be precise, we are located exactly where Voyager 1 had reached in February 1990, a little over twelve years after the probe took off from Earth. We'll put a location map on the viewing wall.

View 42

**Voyager 1
location
1990**

To see outside the ship, use the QR code or
visit www.interstellartours.co.uk/view42.html

It was at this point in its journey into outer space that Voyager
1 undertook a quixotic venture, suggested a decade earlier by
science populariser Carl Sagan. As we have seen, Sagan was
involved in devising the content of both the Pioneer and Voyager
messages. Sagan was an expert communicator and thought that
the project could be used to make a useful point if the probe
was turned around to take a picture of Earth from that distance,
putting our planet's position in the solar system into perspective. There were considerable worries at the time that this could
damage Voyager's equipment if its camera's sensors were over-
whelmed by the light from the Sun – but by 1990, Voyager 1 was
already well beyond the scope of its main mission of studying
the Jupiter, Saturn and Titan, and it was decided that the resultant publicity would be worth the risk.

Voyager's camera was extremely low-resolution even by the
standards of the 1990s. Bear in mind that it had been designed
in the early 1970s, only a few years after the first ever digital
image sensor was developed at Bell Labs in the United States in
1969. It wasn't until twelve years after Voyager 1 was launched
that commercial digital cameras became available, and they
would not become mass-market devices until the late 1990s.

As a result of the poor quality of the original image, we are going to use a filter on our imaging to bring it down to the level that the Voyager camera produced, although the appearance is enhanced compared with the original.

View 43

Pale Blue Dot

To see outside the ship, use the QR code or visit www.interstellartours.co.uk/view43.html

The bands that you can see running from top to bottom in the image are not the ring of a planet but were originally produced by sunlight coming into the camera's lens at an angle and reflecting onto the sensors. (We are simulating this on the viewing wall.) However, there is one tiny bright dot in the most obvious band – what Sagan would give the catchy title of the 'Pale Blue Dot'. It's not a very accurate name. It's hard to see the dot as blue – the colour was more down to expectation than reality, as it occupied less than a single pixel in the original image. But that dot is the far distant Earth, located around 6 billion kilometres (3.7 billion miles) from our current location.

We're going to take a short jump now, past the orbits of the ice giant planets Neptune and Uranus, to take in a spot of sightseeing, with the first stop in orbit around one of the great moons of Saturn.

Running rings around a planet

Once we arrive at Saturn, the view will be dominated by Saturn's stunning rings. Ever since Galileo looked at Saturn through a telescope and drew it with jug ears, we've known there was something special about this planet. Of course, now we are aware that other planets in the solar system have rings – and beyond the solar system there are planets with ring structures that make Saturn look small fry. As we're not quite at Saturn yet, let's take a look at an image of the rings of the extra-solar planet J1407b.

View 44

J1407b:
Super Saturn

To see outside the ship, use the QR code or
visit www.interstellartours.co.uk/view44.html

You can see the star J1407 as a bright flare at the top left, totally eclipsed by those enormous rings. This is a Sun-like star located around 430 light years from Earth. J1407b is a borderline object – it's big for a gas giant planet, but small for a star, so gets labelled both as a kind of supersized Saturn and as a brown dwarf star, the type of cosmic body that's just too small to fully get going with nuclear fusion.

But what J1407b makes up in scale, Saturn is able to eclipse with its sheer elegance. We're now ideally placed to see the solar system's most visually remarkable planet.

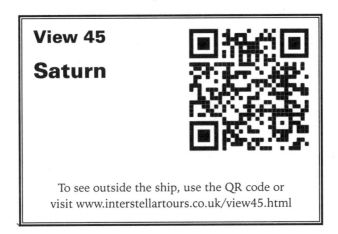

View 45

Saturn

To see outside the ship, use the QR code or
visit www.interstellartours.co.uk/view45.html

Once historical telescopes had been improved sufficiently to
realise that Saturn did not have ears, but rather was surrounded
by what appeared to be a series of concentric rings, there was
uncertainty about how such massive, unsupported structures
could be made up. They were a mystery. It wasn't until 1855 that
the nature of the rings of Saturn was clarified, when the rings
were the subject of a prize competition, entered by the then
newly appointed, 24-year-old professor of natural philosophy at
Marischal College, Aberdeen, one James Clerk Maxwell.

Ironically, given this achievement, Maxwell never showed
any great interest in astronomy. His later work, which put him
in the top ten of the greatest physicists who ever lived, was on
the statistical analysis of the behaviour of gases, and particularly
on the theory of electromagnetism, which led him to believe that
light was an electromagnetic wave. This was the work that led
directly to Einstein's special theory of relativity. But in 1855,
the Adams Prize, a heavy-duty mathematical challenge set by
St John's College, Cambridge, was to present the daunting task
of deciding whether the rings of Saturn were rigid, fluid or made
up of 'masses of matter not mutually coherent'. Making an entry
to the Adams Prize, which is still running today, was and is a
major undertaking.

For Maxwell, the effort was worthwhile as Saturn's rings presented an interesting mathematical challenge. The initially popular assumption that the rings were solid had already been dismissed in 1787 by the French mathematician Pierre-Simon Laplace and Maxwell agreed with this. He pointed out that if, for example, the rings had been made of iron, the gravitational force on such an unsupported structure would have been so strong that they would be ripped apart, even liquifying in parts. Laplace had suggested that it would be possible to make a solid ring stable by having more of the mass located on one side of the ring, but Maxwell calculated that this would only work if 80 per cent of the mass were in a concentrated lump, like a ring with a massive jewel. This clearly wasn't the case for the elegant symmetry of the real rings of Saturn.

Maxwell struggled with modelling fluid rings – the dynamics of fluids are notoriously complex to calculate – but when he found a mathematical method that would work, it became clear that such rings would be unstable thanks to the gravitational impact of Saturn's moons and of its nearest massive neighbour, Jupiter. But his last set of calculations, based on 'an indefinite number of unconnected particles revolving around the planet' worked. Maxwell won the prize for his work – which was not entirely surprising as his was the only entry that year – and it proved the starting point of the realisation that the rings of Saturn are orbiting chunks of rubble, much of it made up of water ice.

It was suggested for some time that the rings could have been the remains of a shattered moon, but it seems far more likely that they were left over material from the formation of the solar system, just as the asteroid belt is left over rubble that never formed into a planet.

It could be life, but not as we know it

Now that we are in the vicinity of Saturn, we can take a closer look at a most remarkable moon, which has for a long time been considered one of the most likely locations for life in the solar system outside of the Earth. This is Titan.

View 46

Titan

To see outside the ship, use the QR code or visit www.interstellartours.co.uk/view46.html

It's immediately clear that this monster of a moon is very different from the Earth's natural satellite. Firstly, it's bigger. Our Moon is unusually large for the size of the planet, but Titan is half as big again in diameter. However, the thing that really makes Titan stand out from other moons of the solar system is that it is one of the very few bodies we know that has a proper atmosphere – it's this that gives Titan its distinctive yellow colouration.

From the point of view of our kind of life, the thick gaseous layer lacks the key ingredient of the Earth's atmosphere. Both are dominated by nitrogen – around 78 per cent in our case and a massive 97 per cent for Titan – but most life on Earth, including us, is dependent on the 21 per cent oxygen content, while

Titan's atmosphere is pretty much oxygen-free. Instead, its other primary constituent (although this amounts to only around 2.7 per cent of what's there) is methane.

If we switch to infrared viewing, we can see through the atmosphere to Titan's surface beneath. The topmost image is the live view of Titan in infrared right now, but we've added in earlier views taken from different directions and the visible light view to get an overview of the moon's appearance.

View 47

Titan infrared

To see outside the ship, use the QR code or visit www.interstellartours.co.uk/view47.html

Clearly, this moon has features that could be interpreted as continents and lakes, but there's a problem with this. Despite the greenhouse effect of Titan's thick atmosphere (it has nearly half as high a pressure again as Earth's atmosphere), the surface of Titan is distinctly chilly at around –180 °C (–290 °F). Much of the surface is frozen, but it does feature a number of lakes. At this temperature, these are clearly not full of water, but rather of liquid methane (and the related hydrocarbon ethane), which is liquid in a relatively tight range of temperatures between –160°C and –182°C (–259°F and –296°F).

Although we know that bacterial life can exist without oxygen, and at wide extremes of temperature, life does require liquids to enable its microscopic mechanisms to function, and some scientists have speculated, since details of Titan were known in the early 21st century, that the moon is a possible home for extra-terrestrial life. The lakes are not the only possibility for life signs to be found – there is also thought to be either liquid water or ammonia underneath the ice in which the lakes reside.

Given the length of time that Titan has been observed, you might think that we could say for certain whether there was life on this moon of Saturn by now. However, for a long time, extreme caution has been observed, as any contamination from Earth could both mislead probes as to what's present and cause damage to Titan's environment. As a result, landings and exploration have been minimised and we are yet to have any definitive direct evidence of living organisms.

A spot of chaos

Leaving Saturn behind, we're taking a relatively short hop to orbit the gravitationally dominant planet of the solar system, Jupiter, where once again it will be a moon that proves of particular interest.

Jupiter is big. It's the next best thing we have to the brown dwarf we saw with those amazing rings – also a failed star, but less massive and so less close to the possibility of fusion occurring. As a primarily gaseous planet, it only has about a quarter of the density of the Earth (its density is quite similar to that of water), but Jupiter is so chunky that its mass is around 2.5 times that of every other planet in the solar system combined. That means that gravitationally, Jupiter has an influence on the orbits of other planets second only to the Sun.

We've positioned ourselves on arrival at Jupiter to see its best-known feature – the great red spot. Note, by the way, the bright dot you can see to the left of the planet, this is one of Jupiter's many moons, Europa, the smallest of the four moons discovered by Galileo, which we will move closer to as soon as you've had a chance to take in the planet itself.

To see outside the ship, use the QR code or
visit www.interstellartours.co.uk/view48.html

When it was first observed, the great red spot was thought to be a physical feature on the planet's surface, but we now know (as is visible from this range) that the apparent surface feature is a roiling mass of gas. It seemed, then, that the red spot was a vast storm like a super-sized hurricane. Because of the scale of the planet, it's hard to really appreciate just how big this feature is. The spot has a larger diameter than the whole of the Earth, although it has shrunk a little in the past century or so.

But there's a problem with the simplistic picture of the spot as an ordinary storm. We are familiar with the nature of storms from those back on Earth, and although Jupiter's atmosphere is very different from our own, even so, we would not expect that storms would remain in place for hundreds of years, as the red spot has done since it was first observed. There's some uncertainty over exactly when this was. It was definitively pointed out by an amateur astronomer, Samuel Schwabe in 1831, but the Italian astronomer Gian Cassini described something similar in 1665. However long it has been there, the answer to this oddity is a mathematical concept given the dangerous sounding name of chaos.

Many natural systems, like the state of Jupiter's atmosphere, are mathematically described as being chaotic. This doesn't mean that they are totally random – quite the opposite. Although there do appear to be some aspects of nature that are genuinely random, notably the behaviour of quantum particles, chaotic systems are deterministic. This means that if you knew everything about them down to the finest detail, you could predict how they would change in the future for ever. But the devil in chaos is in the detail.

Such systems are susceptible to very small changes in the way things start out causing dramatic deviations in how the system develops over time. Because we can never know the exact starting condition of any complex system, both in the number of parameters to measure and the impossibility of having totally exact measurements, it is impractical to predict the behaviour of a chaotic system for too long into the future. There will always be a timescale over which what happens will vary dramatically from expectations – sometimes very suddenly.

To see how small these differences can be, the whole concept of mathematical chaos emerged when American meteorologist Edward Lorenz was working on some of the first computerised weather models in 1961. His computer was slow, taking a long time to complete its calculations. As a result, when he wanted to extend its run, rather than starting the whole process from scratch, he used values from a printout taken partway through the previous run. The result was a forecast that rapidly diverged wildly from the original one. The reason proved to be that the computer was working with numbers to six decimal places, such as 2.591431, but the printout showed the number rounded to three decimal places, which would be 2.591 in that example. It seems an entirely trivial difference, yet it was enough to totally change the way the model's predictions evolved through time.

The wisdom of ensembles

We tend not to be taught about such uncertainty in school science (perhaps the teachers don't like the idea of introducing chaos to their classrooms) – and even at university, it rarely makes much of an appearance in undergraduate work. But in the real world, uncertainty and chaos are commonplace. They arguably represent the norm. Many of the large-scale systems we are familiar with are mathematically chaotic. Think, for example, of the weather and the climate, or of the economy or the progress of a pandemic. This doesn't mean that nothing can be predicted in such circumstances, but rather that we have to take a different approach to using traditional mathematical models.

Until the 1980s, making a weather forecast involved running a computer model of the weather systems and predicting what would happen. But this can often go wrong, due to that dependence on the exact detail of initial conditions. Since then, weather forecasts have been changed to use a technique known as ensemble forecasts. Here, the model is run many times (typically, at least 50 iterations) with very slight differences in the starting parameters of the model. Because we can never know *exactly* what all the details of the real-world environment are, this helps deal with the uncertainty. As a result, ensembles make it possible to produce much better weather forecasts for up to about two weeks ahead, with probabilities of key factors like rainfall dependent on how many of the model runs in the ensemble featured it.

It took far longer for ensemble forecasting (and good modelling) to be introduced to economics, which was based on single-run, unscientific over-simplistic models as late as the 2020s, even though it was clear by then that economies were just as much chaotic systems as the weather. But since then, ensemble forecasting has transformed economic prediction too, and we are now far more likely to have warning of the risk of financial crashes and other economic instability. But it can never be a matter of certainty – predictions of chaotic systems can only ever be probabilistic. Some sudden and unpredicted events will always happen in a chaotic system.

Speculation alert

In 2023, it was still the case that most economists did not believe there was much wrong with their models and would not accept the scientific approach used by meteorologists and others. Back then, economics was still more art than science, but it seems likely that they would have eventually given in to reason.

Limited predictability, improved by using ensembles, is not all there is to chaos, though. It also leads to self-patterning systems. Imagine, for instance, trickling hot water down a sloping board covered in wax. Initially, the water will run over the surface in a chaotic fashion, influenced by tiny, unpredictable variations in the surface of the wax. But over time, parts of the surface will melt. When they do, more water will run down those channels, making them deeper until the water develops a pattern in the surface like a series of tree branches.

There was no initial plan to this pattern, no way to predict exactly how it would emerge. But it is natural in a chaotic system for stable patterns to emerge in this way. And that seems to be the case also with Jupiter's great red spot. It is an unusual form of storm that represents an island of calm in the otherwise regularly changing atmospheric patterns that has spontaneously organised from chaos. The emergence of islands of calm is a typical feature of chaotic systems.

In fact, we do have related phenomena in the Earth's weather systems – it's just that the great red spot's visual resemblance to a hurricane makes it easy to misunderstand its nature. We have also got atmospheric and oceanic processes that have continued for centuries, notably the flows of air in the atmosphere known as the Jet Stream and oceanic 'conveyors' such as the Gulf Stream, which carries warm water up to moderate the climate in parts of Europe. Like the great red spot, their continued existence represents a long-term calm in what is otherwise a more explicitly chaotic system. Equally, like the red spot, they will not last in their current form for ever.

Beneath the ice

Let's move in closer to Europa. This is the least massive of Jupiter's four major moons discovered by Galileo (there are over 70 more moons in orbit) – but despite its relative smallness, it is only just smaller than the Earth's moon. At first sight, Europa has an odd surface that looks a little like the photographs optometrists take of the retinas of human eyes.

View 49

Europa

To see outside the ship, use the QR code or visit www.interstellartours.co.uk/view49.html

The darkest parts are cracks in the surface, with a clear, brighter impact crater visible among them. That surface is otherwise remarkably smooth, as Europa is entirely covered in ice. It has nothing in the way of mountains or valleys. But unlike Titan, what we have here is a predominance of frozen water. Not only is that smooth surface water ice, but beneath it there is a vast liquid ocean. Between them, they make up a covering of water that's around 100 kilometres (62 miles) deep. To put that into context, the average depth of the Earth's oceans is 3.7 kilometres (2.3 miles).

It might seem strange that there would be liquid beneath the outer layer of ice, as the surface temperature on Europa comes close to Titan's frigid lows at around –170°C (–274°F). But Jupiter does not leave its moons to their own devices. Its massive gravitational field creates powerful tides within the aquatic sheath that surrounds Europa.

We are used to tides as an effect that the Moon has on the Earth's oceans – but it is a two-way process. Because the Earth's gravity is six times stronger than the Moon's, its tidal pull on the Moon is six times as strong, and this results in movements in the Moon's surface. It is also why the Moon always presents the same face to the Earth – the tide makes the closest side of the Moon bulge outwards, and over time, the pull of Earth's gravitation on this bulge forced the Moon's rotation into synchronisation with its orbit around the Earth in a process known as tidal locking. Magnify that influence up again, as Jupiter's mass is over 300 times that of Earth, and although Europa orbits somewhat further away from Jupiter than the Moon does from Earth, Jupiter's gravitational field has a much greater impact.

The stretching and flexing that the tides produce on the moon heats up the outer material of Europa, and this is sufficient to liquify much of the water beneath Europa's icy outer surface. (This effect is even stronger on Jupiter's closest large moon, Io, which is not icy and as a result experiences dramatic volcanic activity.) Tidal forces also seem to be responsible for those dramatic large-scale cracks in the surface of Europa.

Because Europa's ocean is so deep and covers its whole surface, despite the moon being far smaller than the Earth, the water layer contains more than twice as much liquid as the entirety of the Earth's oceans. This water is not pure but contains minerals, as evidenced by the dark streaks where the surface ice has cracked.

Here, at last, we have discovered life away from Earth. The warming effect of the tides has raised the ocean sufficiently to become liquid (although not much above melting point for pure water), while there are also thermal vents on the seabed, similar to the so-called black smoker vents found on the bed of the deeper parts of the Earth's oceans. Although the first expeditions to Europa found no signs of life, we have now discovered very basic lifeforms.

Unexpectedly, however, this life appears to be directly related to life on Earth, with DNA and genetic similarities to early bacteria. It has been speculated that life on Earth and on Europa both originated from the same source and that it has been spread by impacts from meteors and asteroids.

It's a rare occurrence, but when a heavy body hits a planet or a moon from space, it can dislodge material so violently that matter from the surface is blasted back out into space. For example, a number of meteorites that have landed on Earth were originally part of the Martian surface, displaced this way. It seems probable that this has happened at some point on the location that was the single, original source of life.

> ### Speculation alert
> As of 2023, no life had been discovered on Europa, although the conditions seem relatively favourable. Europa's liquid water seemed a better bet than the more speculative possibility of life in Titan's hydrocarbon oceans. The idea that life on Earth was seeded from elsewhere in the solar system, known as panspermia, was supported by some leading scientists, but was not widely held. However, if such an event were to occur, it is unlikely that life travelled from Europa to Earth, and more likely it was in the other direction, or that both originated from a third source.

The outer neighbour

Leaving Europa behind, we are making the jump across the dividing line of the solar system marked out by the asteroid belt. As we saw when visiting a protoplanetary disc, this isn't, as was long thought, the remains of a shattered planet, but is simply left-over debris from the Sun's early protoplanetary disc where there was insufficient matter to make up a planet, were it ever assembled.

The asteroid belt is no obstacle to us, thanks to hyperspace, but even if we were an old-fashioned rocket, cruising through conventional space, it wouldn't be the kind of dire navigational threat that is portrayed in space-based adventures. There, tumbling asteroids form a complex, three-dimensional maze that ships have to fight their way through, providing a huge challenge to the pilot's skills. This has made for great spin-off computer games, but in reality, getting through an asteroid belt is a piece of cake.

In fact, if we were to come out of hyperdrive at a random location in the asteroid belt – it would be pointless as part of our tour because, almost certainly, all you would see would be the blackness of space and the distant stars – there would be nothing nearby. The average distance between bodies in our asteroid belt is about 1 million kilometres (620,000 miles). There's plenty of open space.

When we return to normal space next, in Mars orbit, we will have moved from the gas and ice neighbourhood of the outer planets to the rocky locations that are more familiar from our experience of Earth. And few planets have featured so thoroughly in our imagination as possible sites of extra-terrestrial life than Mars. In fact, in the early to mid-20th century, the concept of aliens was pretty much synonymous with Martians, thanks in part to H.G. Wells' classic novel *The War of the Worlds* and its enigmatic Martian invaders.

Initially, Mars' appeal as a planet was due to its distinctive appearance, showing up as clearly red in the sky – hence its name that links it to the Roman god of war. But as naked-eye astronomy was superseded by use of telescopes, it became clear that Mars had some similarities to the Earth in having visible ice caps and surface features, some of which changed with the seasons, suggesting perhaps that there was vegetation. There was also a brief period of excitement when it was thought that clear evidence for intelligent life on Mars had been uncovered.

This started back in 1877 when the Italian astronomer Giovanni Schiaparelli took the opportunity of Mars being at its closest point to Earth* to draw detailed maps of the planet's surface. In retrospect, these maps were far too detailed for

---

* The orbit of Mars is significantly more elliptical than Earth's, and as a result of this and different orbital periods, the distance between Earth and Mars varies widely, between around 56 million kilometres (34 million miles) at the closest and 400 million kilometres (249 million miles) at the most distant.

the resolution of the technology available – Schiaparelli was seeing detail that his telescope was simply not capable of rendering; details that he identified as a natural network of water channels, crossing the planet from the poles.

Although these channels aren't really there on the surface, on their own, Schiaparelli's observations were not a huge error. It certainly wasn't the first time an astronomer's imagination filled in detail that proved incorrect. However, a dubious translation from the original Italian 'canali' to English resulted in the imagined features being described as 'canals'. It seemed as if Schiaparelli was drawing a planet-wide, Martian transport system constructed by intelligent lifeforms. This idea was taken up by the otherwise respectable American astronomer Percival Lowell, who produced even more detailed maps of these imaginary features in the early 1890s, culminating in the publication in 1895 of a book called *Mars* describing them in enthusiastic detail.

Here, Lowell waxed lyrical, commenting: 'Quite possibly, such Martian folk are possessed of inventions of which we have not dreamed ... what we see hints at the existence of beings who are in advance of, not behind us, in the journey of life.' Two years later, H.G. Wells opened *The War of the Worlds* with the chilling words:

> No one would have believed, in the last years of the 19th century, that human affairs were being watched keenly and closely by intelligences greater than man's and yet as mortal as his own; that as men busied themselves about their affairs they were scrutinized and studied, perhaps almost as narrowly as a man with a microscope might scrutinize the transient creatures that swarm and multiply in a drop of water.

A little over 70 years later, probes began to regularly be sent into Martian orbit and would eventually land on the planet.

This was despite something of a litany of failure, in part caused by the delay in communication time to Mars that we've already explored. The robotic visitors found dramatic scenery, including the huge Olympus Mons mountain, a long-extinct volcano that is 22 kilometres (13.6 miles) in height. They also found signs of open water having once been present in dried-up water courses. But there was no life.

Mission to Mars

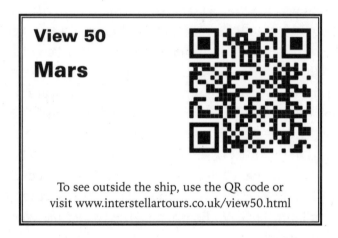

View 50

Mars

To see outside the ship, use the QR code or visit www.interstellartours.co.uk/view50.html

Although there are some visual similarities to Earth, the more we have discovered about Mars, the less likely that life on the planet seemed possible outside of a David Bowie song or a Ray Bradbury story. The planet is considerably smaller than the Earth, with about 40 per cent of the strength of gravity. It receives less than half the amount of energy from the Sun than we do at home and has a very thin atmosphere with practically no oxygen in it. There seems to have been a thicker atmosphere in the past, but Mars does not have a significant magnetic field (unlike the Earth), which means that it is not protected from the solar wind that tends to strip off atmospheric gas.

While it's true that Mars does have those impressive ice caps, what is on show is frozen carbon dioxide – dry ice. There is plenty of water ice on Mars – in fact, most of the polar ice under that carbon dioxide frosting is water – but the ice is found mostly in out-of-the-way places. The temperature on the surface of the planet ranges from a minimum of –110°C (–166°F) to a balmy 35°C (95°F), but the average of –60°C (–80°F) makes it clear that most of the time any water present will be frozen.

For a long time, Mars was a popular destination for science fiction films and books before the sheer difference in scale of a journey to the Moon and to Mars really began to sink in. A trip to the Moon from Earth in an old-fashioned rocket took about three days – but using that technology to get to Mars used to take between six and nine months. That's not just an issue in terms of the elapsed time that the astronauts had to be stuck in the confines of a small spaceship. It meant carrying a huge amount of food and drink, along with needing to protect the astronauts from the impact of the radiation from the Sun over such a long period.

Nonetheless, Mars was first successfully colonised in the 2060s, and since then, efforts have been made to make the planet closer to something inhabitable by humans, a process known as terraforming. The Martian atmosphere has been thickened and plants are increasingly being grown after some initial work using robots to break up the soil and prepare the location for the second wave of colonists. It remains an extremely harsh environment, though.

Speculation alert
Back in 2023, no human had ever set foot on Mars, and getting there was a far bigger challenge than getting to the Moon, both in length of flight (averaging out at around 70 times as long) and in the implications of such a long journey in terms of both

provisioning and keeping astronauts safe from radiation and technical failures. However, even then it seemed a reasonable speculation that despite the limitations of the technology of the 2020s, a colony would be set up on Mars within twenty to 50 years. As described below, robotic plants to make the Martian surface more habitable had already been proposed in the 2020s.

Back in the early 21st century, Italian scientists had proposed to the European Space Agency the idea of 'plantoids', robots based on plants that would produce energy from sunlight using solar panel leaves and that would have roots that extended into the soil of Mars to break it up, search out water and produce the materials that would be needed for Martian buildings to be constructed.

This was a real breakthrough in thinking. Space designers are primarily engineers and tend to think of engineering solutions to problems. This meant that, for example, when they sent a probe to the comet 67P/Churyumov-Gerasimenko in 2014, it was a near disaster. The probe, called Philae, could not be controlled remotely. Signals from Earth took around half an hour to reach Philae's parent ship, Rosetta, and frequently Rosetta could not communicate with Philae because the comet was in the way. Philae was supposed to anchor itself to the comet using harpoons, but on landing, it bounced and the harpoons didn't work. As a result, the probe bounced several times more, ending up partially in the shade and unable to get enough electrical charge to carry out much of its mission.

Unlike harpoons, though, plants take a much more subtle route to getting anchored in even the hardest of soils. The plantoids were dispersed across large areas of Martian landscape and not only broke up the surface, but by producing raw materials made it far quicker for colonists to be able to settle. You may be surprised, though, if you haven't been following the Mars colonisation that there is no sign of either habitats or farms on the surface. Some of the dark areas visible there are vast fields of solar panels. And if we zoomed in close enough, you would be able to see a couple of ships currently on Mars. But the vast majority of the Martian colony's presence is underground.

We've already mentioned how Mars lacks a magnetic field and an atmosphere. On Earth, these, combined with the ozone layer in the upper atmosphere, help protect us from excessive radiation. We are totally reliant on the Sun's energy, but our neighbourhood star also pumps out high-energy light and physical particles in the so-called solar wind that can do plenty of damage to living things.

On Earth, we're familiar with having to wear sunscreen to keep ultraviolet from causing skin damage and potentially harming our cells sufficiently to cause cancer – but things are far worse without the Earth's array of protective shields. Mars is considerably further from the Sun, but, until the terraforming process manages to significantly change the atmosphere, it is unprotected from the direct impact of solar radiation – and that means that the surface is constantly bombarded by high-energy photons that cause cancer in humans and genetic damage to plants.

Despite Mars getting plenty of sunlight, the job of those plantoids was not to make it possible to plant normal crops, but to break up the ground for future excavations and to provide material for the 3D printers that have constructed the buildings and much more for the colonists. The people live underground, and their farming is hydroponic, using water, nutrients and artificial light. Mars still provides some of the requirements for this. There's plenty of energy from solar, water and oxygen from ice, and carbon dioxide from the atmosphere, but the surface of Mars is not going to be habitable any time soon.

From here, it's not far (in interstellar terms) to reach home.

Send in the clouds

You might think we've got back to Earth on an exceptionally cloudy day, but we have intentionally overshot our final destination. What you can see on the viewing wall now is the planet Venus.

View 51

Venus

To see outside the ship, use the QR code or
visit www.interstellartours.co.uk/view51.html

Venus is Earth's other nearest neighbour. Like Mars, it has attracted attention since prehistory, in this case not for its colour but because it's the brightest thing in the night sky after the Moon. Soon after darkness falls, or shortly before the dawn, when Venus is in a clear sky, it's an unmistakable sight. In some historical periods, it was thought to be two different bodies, which formed the morning and evening star respectively. Because Venus is closer to the Sun than Earth is, it is always relatively close to the Sun in the sky, so is only ever seen for a couple of hours after sunset or before the dawn.

Because Venus was such a striking feature of the night sky, as telescopes came into use, astronomers were excited to see more of what the planet could offer – but they were soon to be disappointed. However good their instruments, Venus remained veiled, a near-featureless, bright white blob. Admittedly, the surface did appear to have faint markings, which some initially interpreted as land features. Fresh from his Martian 'discoveries', Percival Lowell somehow picked out distinct straight-line land markings, but unlike his canals, no one else seemed capable of reproducing these.

The Greek–French astronomer Eugène Antoniadi commented on Lowell's alleged discovery: 'Forgetting that Venus is decently clad in a dense atmospheric mantle, he covered what [he] called the "surface" of the unfortunate planet with the fashionable canal network, dividing it into clumsy melon slices.' It was eventually proved that Venus had indeed such heavy cloud cover that its surface was never visible from Earth. Those faint markings, which weren't straight lines, were the result of pollution in the atmospheric makeup that changed over time.

Even so, as it became clear that Mars was unlikely to house higher forms of life, Venus started to be seen as the best possibility for a second Earth. Admittedly, it was obvious that it was going to be hotter than home – Venus is a bit less than three quarters of the distance from the Sun that the Earth is. However, in other ways, it's significantly more like the Earth than Mars. To begin with, Venus is much closer in size, experiencing about 0.9 times Earth gravity – low enough to feel pleasant, but not so low that it feels odd. And with that cloud cover, it was more likely to be screened from the worst treatment that the Sun could dole out.

For decades, science fiction writers portrayed Venus as a hot, sticky jungle planet – but one that was covered with an array of flourishing, active life, rather like the then imagined foliage from the time of the dinosaurs. It was something of a shock when the first successful Venus probe, NASA's Mariner 2, reached the planet in 1962 and started to send back data that uncovered the reality of this hellish planet. The surface temperatures on Venus averages around 460°C (860°F) and can reach as high as 600°C (1,110°F). This makes it significantly hotter than Sun-hugging innermost planet Mercury. The metal lead would run liquid on the surface of Venus.

Inside the Venusian greenhouse

The planet does have a dense atmosphere, one of the reasons it was thought to be a hopeful home for life, and perhaps for human colonisation – but it is nothing like our own atmosphere. Venus is swathed in an abundantly thick layer of carbon dioxide, giving it an immense atmospheric pressure at the planet's surface. This is over 90 times the pressure we experience on Earth. The over-the-top greenhouse effect that all the carbon dioxide in the Venusian atmosphere produces is responsible for those incredibly high temperatures.

We tend to think of the greenhouse effect as being bad because of the impact it had on the Earth's global warming before it was brought under control and reduced in the second half of the 21st century. In moderation, though, the greenhouse effect is highly beneficial to life back home. If there were no greenhouse effect at all on Earth, temperatures would be about 30 degrees lower, averaging –18°C. This would mean that there would be very little that could survive, except specialist bacteria that thrive on extreme temperatures (though these probably would never have originated). However, too much of the greenhouse effect certainly is a bad thing, as Venus amply demonstrates.

We never see any of the features of Venus from orbit, the way you can when looking at the Earth from space despite our cloud cover. This, as we've seen, is because the Venusian clouds are so thick, providing a permanent near-white screen to what's below. Although the atmosphere is primarily carbon dioxide, it's not the CO_2 that we see, though – carbon dioxide is transparent. As if to emphasise just how hellish the environment of Venus is, those bright white clouds that totally conceal the surface and make the planet so bright are the intensely vitriolic substance sulfuric acid*. With such thick covering, despite Venus being nearer to the Sun than Earth, less light gets through to the planet's surface in the immensely long Venusian daytime.

It's a strange day down there because of the way that Venus rotates, which is in the opposite direction to the Earth and every other planet in the solar system except Uranus. The rest of the planets rotate anticlockwise as seen from their north poles. This is the same direction in which the Sun also rotates – and follows the direction that all the planets orbit the Sun. It's natural for rotations and orbits to be in the same direction as they would all have been initiated by the rotation of the original protoplanetary disc.

In the case of Uranus, this 'retrograde' rotation is likely to have been the result of a collision with another large body when the solar system was first forming, but Venus seems to have developed its odd behaviour in a different way. The Sun's tidal impact on Venus would have slowed it down considerably sooner than the equivalent effect of slowing the rotation speed of the more distant Earth. This brought Venus towards tidal locking, as the Moon is locked to present the same face to the Earth. Meanwhile, a second factor would have been that immensely thick atmosphere which has enough of a drag on the

---

* Apologies to any historians of science for what amounts to tautology in saying 'the intensely vitriolic substance sulfuric acid'. Vitriol was an old name for sulfuric acid – but that use is now sufficiently obsolete that the wording makes sense.

surface to overcome the tidal lock and slowly spins the planet in the opposite direction to its original motion.

Due to its sluggish rotation, the day on Venus, in the sense of a complete rotation of the planet, takes 243 (Earth) days to complete. This is longer than the 225 days that Venus takes to get around the Sun. But because of the way that the planet spins combined with its movement around the Sun, the time between sunrises is just under 117 days, with the Sun rising in the west and setting in the east. Working this all out in the early days of telescopic observation was complicated by motion of upper parts of the atmosphere, which rotate in just 4.3 days.

Below those clouds is a post-volcanic landscape, unlike any other planet in the solar system. Apart from very local images taken by landers, which still only tend to survive for hours under Venus' conditions, most of the views we have of the planet's surface are taken by radar. This gives a distorted view of reality because, for example, the dark and light patches refer not to visual brightness but indicate whether the surface is smooth (dark) or rough (bright). We can switch the viewing wall to radar, which takes a while to build an image, but has been working in the background since we arrived.

View 52

Venus radar

To see outside the ship, use the QR code or
visit www.interstellartours.co.uk/view52.html

There are features here such as mountains and craters. In what was arguably a somewhat patronising fashion, in the 1970s, the International Astronomical Union decided to give all features on Venus female names. There is a single exception (made before this decision). A mountain range is called Maxwell Montes after Scottish physicist James Clerk Maxwell as Maxwell devised the electromagnetic theory behind the radar that was used to discover the surface details of the planet.

Now, though, it really is time to go home.

A special place

We're just about to arrive back in Earth orbit, assuming that Earth is still there. After all, it is, strictly possible that Earth has left its orbit since we last saw it from the pale blue dot location on the edge of the solar system. This possibility is a function of a shocking discovery made by Isaac Newton, but one that is largely concealed by the way that we talk about astrophysics. Orbits are not fully predictable – just like Jupiter's great red spot, they have a chaotic element.

When Newton did his work on gravitation, he was able to calculate exactly how the orbit of the Earth around the Sun, or the Moon around the Earth, taken in isolation, would work. This is the precision science of the clockwork universe. It was a picture that so enraptured the French mathematician Pierre-Simon Laplace that he conjured up the idea of what has become known as Laplace's demon. He described this imaginary being as follows:

> An intellect which at any given moment knew all of the forces that animate nature and the mutual positions of the beings

that compose it, if this intellect were vast enough to submit the data to analysis, it could condense into a single formula the movement of the greatest bodies of the universe and that of the lightest atom; for such an intellect nothing could be uncertain and the future just like the past would be present before its eyes.

Before the understanding of uncertainty and chaos we now have, this seemed a perfectly reasonable concept, even if the demon itself was only ever a piece of fiction. But in the real world, as Newton discovered and Laplace must have known, fully predictable determinism has a fatal flaw. As we saw at Jupiter, many natural systems are chaotic.

Systems such as the weather, or the economy, which we have used as models of chaos, have a lot going on in them, so chaos perhaps seems a more likely outcome. But the thing that shocked Newton and his contemporaries was that the behaviour of some very simple systems were also impossible to predict indefinitely into the future, because they too had chaotic elements (although back then they did not understand the nature of chaos in a more general sense).

Specifically, while Newton was able to predict with perfect accuracy the future behaviour of an imaginary system consisting of two astronomical bodies, such as the Sun and the Earth in isolation, as soon as a third body was added, things got messy. Decades turned into centuries as many tried and failed to solve the 'three body problem'. By the 19th century, French mathematician Henri Poincaré was able to prove that it is impossible to produce a formula that fully describes the behaviour of three or more bodies interacting gravitationally.

The approach taken ever since Newton (though refined since) to get an approximate solution is known as perturbation.

An initial calculation is made for just two bodies, and then the effect of the third body added in – but this will always have some limitations in accuracy. The real solar system is far harder to deal with than this because there are far more than three bodies involved. So, for example, when considering the motion of the Moon around the Earth we have to take in the influence of the Sun, Jupiter and more. While it's possible to find a solution to any required degree of accuracy for a fixed length of time, no prediction can describe in detail what will happen for certain and for ever.

Like the models that so improved weather forecasts towards the end of the 20th century, astronomers have now run many ensemble forecasts of the future of the solar system. Although it is impossible to ever say 'never', they have as yet not found a prediction of the Earth disappearing from its orbit and heading off into space. For us, about to re-emerge into real space in Earth orbit, this is a reassuring fact.

Just to show, though, that this doesn't mean that the solar system as a whole will remain stable as it is into the indefinite future*, the ensemble modelling of its interactions gives Mercury a 1 per cent chance of leaving its orbit and colliding with Venus within a few billion years.

Thankfully, though, as we emerge from hyperspace, we do indeed have the beautiful sight of the Earth to greet us.

---

* We know for certain that, in once sense at least, the solar system will not remain as it is indefinitely because the Sun has a known lifespan before it fluffs up as a red giant about 5 billion years in the future, expanding so far that it will engulf Mercury, Venus and probably Earth, before its outer parts are dispersed, leaving a white dwarf. This will certainly increase solar system instability for anyone remaining to watch it.

View 53

Earth

To see outside the ship, use the QR code or
visit www.interstellartours.co.uk/view53.html

Even in context, a special place

Without doubt, our home planet is special within the solar system. Visually, you could argue that Jupiter and Saturn are more dramatic. But in terms of its varied environment and having all the components that give life the ability to thrive (not to mention having all that life existing), it is unique. Earth might have been a pale blue dot from the outer reaches of the solar system, but from orbit we can see it as a richly varied tapestry.

In one way, our trip to see different highlights of the galaxy has made it clear to us just how small and apparently insignificant the Earth is. In his script for the original *The Hitchhiker's Guide to the Galaxy* radio series, Douglas Adams imagined a device known as the total perspective vortex, which drove others mad, but simply allowed the supremely self-important character Zaphod Beeblebrox to realise that he was 'a really great guy'.

When placed in the vortex, according to Adams' script, 'you are given just one momentary glimpse of the size of the entire unimaginable infinity of Creation, along with a tiny little marker saying, "You are here".' While our trip only encompassed the Milky Way as opposed to that 'unimaginable infinity of Creation', we have still put the Earth into context as a single, small location among at least 100 billion stars, many of which have one or more planets. Many of us consider travelling 1,000 kilometres (620 miles) a significant distance. But the Milky Way is around 87,400 light years across. That is 8.3 x 10^{17} kilometres or 5.1 x 10^{17} miles. As perspective goes, our lives and our world are tiny compared to the galaxy as a whole.

Some take this to mean that the Earth is insignificant, but it is not. We don't know if intelligent life is limited to Earth, but we have not encountered any other intelligent forms of life out there, with no known life at all so far discovered outside the solar system. Even if there is life elsewhere, we know that it is a relatively rare occurrence.

In some ways this isn't surprising. We've seen the concept of the Goldilocks zone, with the Earth being the right distance from the Sun to get sufficient energy to keep water liquid without being so hot that it is only present as vapour. But there's significantly more. As we saw with Proxima Centauri b, it's possible for a planet to be in the zone and still be uninhabitable. The Earth has its screens of ozone, the atmosphere and its magnetic field to protect it from the Sun's more malignant output. It has just the right kind of atmosphere.

The Earth also has an unusually large Moon that has helped make the environment relatively stable. There are several benefits we've gained from having our Moon. Some come from the way it was formed. The evidence suggests that before the solar system had settled into stability, the early Earth was impacted by another planet-sized body. Part of the resultant merged body was splashed out to form the Moon. But this was the lighter part of the whole. The iron core of the two bodies (and their associated, even heavier radioactive material) combined to give the Earth an unusually large metal core, which is responsible for our abnormally strong magnetic field.

As we saw when looking at exoplanets, and as also applies to Mars, when a planet doesn't have a strong magnetic field, it makes conditions for life significantly less likely to be retained, as the stellar wind will both strip atmospheres away and batter the surface with radiation. The formation of the Moon had a direct influence on the Earth becoming more life-friendly. What's more, it resulted in our home planet having an unusually thin crust.

This is why the Earth has plate tectonics – the movement of large sections of the crust across the surface of the planet. This gives us a carbon cycle that (without human interaction producing climate change) keeps levels of carbon dioxide in our atmosphere at the right sort of levels to have liquid water on the planet. The large Moon has also helped stabilise the Earth's potential wobbles in its orbit, keeping the planet in an orientation to the Sun that prevents any side of the planet being overly exposed to heat and ultraviolet rays.

Earth has also experienced two highly unlikely events – the initial origin of life and then the development of complex life. Considering the Earth to be special is sometimes described as exceptionalism. To those who use this negative label, it seems that we are mistakenly thinking of humanity as somehow wonderful and in charge of everything. But in reality, the special nature of the Earth should be seen as a warning. As we return to our familiar lives on Earth, it is a reminder that we have a responsibility to look after this special environment where life can flourish to keep it that way.

We have to look after the Earth itself to keep it a great location for life, because looking out at the rest of the solar system, we can see that we have a unique place to live. It's true that we can hope that the colony on Mars will be able to grow, and it may be that at some time in the far distant future, Mars will be more like a second Earth – but for the foreseeable future, it is an outpost with conditions that are extremely hostile to life.

Back in the 21st century, humanity finally realised just how precious and precarious the Earth's environment was. The Earth as a planet will be fine, whatever we humans do. It will recover from anything we can inflict – it has been through worse in its 4.5-billion-year history. But the same is not true of an environment that supports and nourishes life.

During the time that the Earth has existed, it has sometimes been far more hostile to advanced lifeforms than it is at the moment. And as manmade climate change started to take hold, we were at risk of pushing the Earth back into a state that was far less friendly to the continuation of life on the scale to which we had become accustomed.

Things did get uncomfortable for a few decades, but we pulled the Earth back from the brink. In the future, there will be other challenges, whether the result of human action or the physical environment. It could be there will be another asteroid on an impact course, threatening the kind of disaster that wiped out the dinosaurs, or there could be natural changes to the Earth's climate that will once again make it less human-friendly. It is up to us as the stewards of life on Earth to make use of our technological capabilities and imagination to keep the Earth safe for the future of life.

I hope you have enjoyed your tour, experiencing the wonders of our galaxy – and that your perspective on our place in the universe, both as humans and as occupants of Earth, has been enhanced and enriched. It's been a pleasure to have you on board.

APPENDIX: SPECIAL THEORY OF RELATIVITY FOR BEGINNERS

There is no need to read this extra material to enjoy our interstellar tour, but if you were intrigued by the suggestion that the mathematics involved in the special theory of relativity could be followed by a high-school student, I wanted to provide the opportunity for you to see just how the fixed speed of light results in the remarkable time-bending results of Einstein's remarkable work at a level more detailed than the descriptive approach taken in the main part of the guide.

The mystery of equations

There will be equations.

If you have problems with these as a hangover from school days, bear in mind that an equation is just a shorthand that makes it easier to see and manipulate mathematical objects – or in our case, to be able to calculate how physical things will behave. Each of those scarily compact letters in an equation merely stands in for something straightforward. Some of these will be constants – in effect, just useful numbers. So, by using *c*, for instance, we are saved from having to type 'the speed of light in a vacuum' or 299,792,458.

Equally, a letter could stand in for a variable. A variable is a harder concept to get your head around than a constant, but it is extremely useful because it's like a container into which you can put whatever is most relevant to you at the moment. So, for instance, we will often come across *v* – short for velocity. The useful thing about a variable like *v* is that we can pop into it whatever value is relevant to the thing we are studying. If, for example, we're looking at a spaceship travelling at 100,000 kilometres per second, then we can just place 100,000 into *v* and we're away.

The only proviso here is about units – a unit being something like 'kilometres per second', which describes the scale of the numbers being used. It's fine to use any unit you like that fits your requirement, but it is important to be consistent. In the case of velocity, we are dealing with a unit of distance per unit of time. I could equally well use feet per year, although it's not how most of us think of speed. Or, for that matter, I could use your height (whatever it happens to be) per millennium. The only problem is that once I've decided on a distance unit, I need to use the same one for everything in the equation, or else I'm going to have problems.

You can see why this is the case by thinking about speed limits. If the limit on a road is, say, 60, meaning '60 miles per hour', it's no use saying, 'I didn't break the law because I was only doing 2', meaning '2 miles a minute', even though that may truly be the case. You need to use the same units to measure your speed as the limit does – in this case, indicating you were travelling at a more fine-inducing 120 miles per hour.

To avoid having to check the units being used all the time, there are standard units in the scientific system known as MKS, which stands for metres, kilograms and seconds. This means that we would usually expect scientists to measure velocity in metres per second, making the speed of light just under 300,000,000 metres per second.

So now that, hopefully, equations potentially hold less fear for us, let's try out the most famous equation in existence:

$$E = mc^2$$

This has all the basic aspects of an equation in a simple, digestible form. It's an equation because of that equals sign. That tells us that the thing on the left (E) has the same value as the thing on the right (mc^2) – they are equal, hence '*equation*'. We've already met c as a constant: it's the speed of light, but here it is squared – multiplied by itself. And then we've got two variables, energy (E) and mass (m). The neat thing about an equation is that either of these can be given any value we like. We then just plug in the numbers, and we can work out the other one. If we know energy, we can find the equivalent mass. Know the mass and we have a route to discover the available energy. It's as painless as that.

A final, useful extra when dealing with the equations required to see special relativity in action is that we need to differentiate between, for instance, the time on the Earth and the time on a spaceship. It is traditional in books on relativity to indicate these two different 'frames of reference' by having one set of variables that are normal (e.g. v, t) and one set that have a little blip attached (e.g. v' and t'). I think it is much clearer to be explicit with the labels and call them, for instance, t_{Earth} and t_{Ship}. I suspect the reason that academics tend not to do this is partly because they have to deal with much more complex equations than we will be handling, and partly because the 'dash' notation started off on a blackboard, where it would be very difficult to read subscripts like 'Earth' and 'Ship'. We have the luxury of proper typesetting.

Introducing the light clock

As we saw in the main text, our starting point in special relativity is an imaginary device called a light clock. This is just a beam of light travelling up and down between mirrors on a travelling spaceship, with the light shone at right angles to the direction the ship is travelling. From the point of view of an astronaut on the ship, the light travels up and down vertically. But as far as a viewer on Earth is concerned, the light beam travels diagonally at an angle, because the ship will have moved between the light leaving the top of the clock and arriving at the bottom.

All we need to produce the remarkable consequence of time dilation – the idea that time slows down on the moving ship as seen from the Earth – is a mental image of that clock, a spot of basic geometry and the idea that light goes at the same speed however you move.

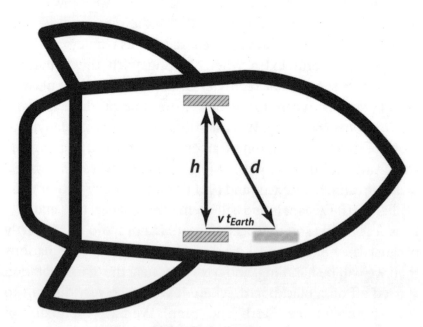

Light clock diagram.

We'll start by looking at the clock from the viewpoint of the astronauts on the ship. Let's say that the height of the clock is h. The speed of light is always referred to as c (because it's constant). So, the distance the light travels, h, will be t_{Ship} / c where t_{Ship} is the time it takes for the light to get from one mirror to the other, as measured on the ship.

Now let's jump back from the ship to Earth and watch that same event. We see the light beam taking a diagonal course because the ship is moving. Thanks to Pythagoras' theorem, we can work out the distance the light travels as seen from Earth. Let's call it d. The other bit of our triangle is the distance the ship has moved forwards. That's the velocity of the ship, v, multiplied by the time that has elapsed from our earthly viewpoint, t_{Earth}. To keep our equations simple, we're going to use the convention that we don't need to put in a multiply sign, so v times t_{Earth} is just vt_{Earth}.

Digging back to school geometry, Pythagoras tells us that $d^2 = h^2 + (vt_{Earth})^2$.

Given that light always travels at the same speed, we know that the distance d is just ct_{Earth} – the speed of light times the time it took. So that makes the equation:

$$(ct_{Earth})^2 = h^2 + (vt_{Earth})^2$$

We'll take away $(vt_{Earth})^2$ from both sides:

$$(ct_{Earth})^2 - (vt_{Earth})^2 = h^2$$

... which is the same as:

$$t_{Earth}^2 (c^2 - v^2) = h^2$$

Now we divide both sides by the bit in brackets:

$$t_{Earth}^2 = h^2/(c^2 - v^2)$$

We're almost there. A little further up we said that h was t_{Ship}/c, so let's stick that in:

$$t_{Earth}^2 = t_{Ship}^2/(c^2/c^2 - v^2/c^2)$$

which is:

$$t_{Earth}^2 = t_{Ship}^2/(1 - v^2/c^2)$$

Finally, we take the square root:

$$t_{Earth} = t_{Ship}/(1 - v^2/c^2)^{1/2}$$

... where that $^{1/2}$ means the square root of what's in the brackets. And that's it.

We've just shown how time dilation emerges from the simple assumption that light travels at the same speed however you move. Because t_{Ship} is being divided by a number that is smaller than 1, more time will have passed on Earth than passes on the ship – in effect, the time on the ship is running more slowly. When v is close to c, the number t_{Ship} is divided by is much smaller than 1, so the time that passes on Earth is much longer than that in the ship.

The formula above is the real thing, the time dilation formula from special relativity, reached with nothing more than Pythagoras' theorem and a spot of algebra.

Discovering time travel

Now, admittedly, if you were never very comfortable with algebra, what we just did was a little messy and difficult to get your brain around. However, there is absolutely nothing there that a sixteen-year-old who can cope with maths in high school would not be comfortable with – in UK terms, this is straightforward GCSE maths – and it took less than a page to come up with the result. This isn't, as it happens, how Einstein himself came up with the relationship. However, it is just as legitimate an approach and it demonstrates why it is bizarre that we don't cover relativity at school.

Now we can immediately see how travelling quickly enables a form of time travel by plugging a speed for the ship into the equation. Let's say the ship is going at a respectable half of the speed of light. So, our time dilation equation gives us:

$$t_{Earth} = t_{Ship} / (1 - (c/2)^2 / c^2)^{1/2}$$

Which is:

$$t_{Earth} = t_{Ship} / (1 - c^2/4c^2)^{1/2}$$

Let's cancel out the c^2:

$$t_{Earth} = t_{Ship} / (1 - 1/4)^{1/2}$$
or
$$t_{Earth} = t_{Ship} / (3/4)^{1/2}$$

The square root of ¾ is around 0.866 so:

$$t_{Earth} = t_{Ship} / 0.866$$

Which is the same as:

$$t_{Earth} = 1.155 \, t_{Ship}$$

And there is our time travel. If the ship travels for ten years at this speed, when it gets back to Earth, 11.55 years will have gone by. The closer the ship gets to the speed of light, the closer that v^2 / c^2 gets to 1, so the number on the bottom of the equation gets smaller and smaller, making the multiplying factor bigger and bigger. The closer you get to the speed of light, the more time slows down on the ship as seen from Earth – the more the ship travels into the Earth's future.

It is possible to do something very similar with a light clock that is aligned with the direction of travel, rather than at right angles to it, to show how length is contracted for a moving object. The final significant aspect of special relativity, the increase in mass of the travelling ship, is a little more fiddly but perfectly possible by throwing in the conservation of momentum.

It ought to be stressed again that this light clock approach was not the one originally involved in deriving the effects of special relativity. Einstein and his contemporaries built on the impact of the Michelson–Morley experiment of 1887, which showed that two light beams at right angles in a moving frame of reference did not alter as the beams were rotated. The mathematics to explain this, and hence derive the basics of special relativity, is not a whole lot more complex, but it is not as easy to follow as the light clock, which is why I prefer to use this example.

Adapted from The Reality Frame *by Brian Clegg*

For further reading, please visit
www.interstellartours.co.uk